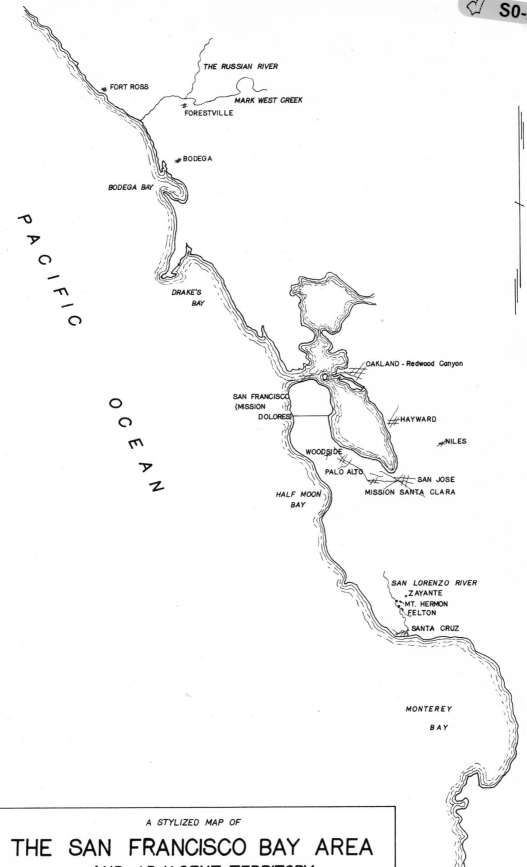

THE RUSSIAN RIVER

FORT ROSS

MARK WEST CREEK

FORESTVILLE

BODEGA

BODEGA BAY

PACIFIC

DRAKE'S
BAY

OCEAN

OAKLAND - Redwood Canyon

SAN FRANCISCO
(MISSION
DOLORES)

HAYWARD

NILES

WOODSIDE

PALO ALTO

SAN JOSE

MISSION SANTA CLARA

HALF MOON
BAY

SAN LORENZO RIVER
ZAYANTE
MT. HERMON
FELTON
SANTA CRUZ

MONTEREY

BAY

A STYLIZED MAP OF

# THE SAN FRANCISCO BAY AREA
## AND ADJACENT TERRITORY

drawn by DAVID W. BRAUN                    1974

# LOGGING THE REDWOODS

# LOGGING THE REDWOODS

*By*

LYNWOOD CARRANCO AND JOHN T. LABBE

THE CAXTON PRINTERS, LTD.
Caldwell, Idaho
1975

Library of Congress Cataloging in Publication Data

Carranco, Lynwood.
    Logging the redwoods.

    Includes index.
    1. Redwood.  2. Lumbering—California—History.
3. Lumbering—California—Pictorial works.
I. Labbe, John T., joint author.  II.  Title.
SD397.R3C35  1975        634.9′758′8209794        72-80989
ISBN 0-87004-236-X

Lithographed and bound in the United States of America by
The CAXTON PRINTERS, Ltd.
Caldwell, Idaho 83605
119621

This book is dedicated to the pioneer lumbermen who succeeded in launching careers as mill men by overcoming the tremendous obstacles of moving the giant redwoods from the woods to the mill, by inventing equipment strong enough to handle the gigantic logs, and by finding suitable markets for their lumber throughout the Pacific area; and to Augustus William Ericson and the other early photographers who preserved the early history of logging in pictures.

# TABLE OF CONTENTS

                                                              *Page*

DEDICATION ..................................................  V

LIST OF ILLUSTRATIONS ....................................  ix

*Chapter*

ONE      THE REDWOODS ......................     1

TWO      EARLY SAWMILLS .....................     7

THREE    OPENING THE NORTH COAST ...................    13

FOUR     EARLY LOGGING .............................    19

FIVE     MENDOCINO LANDINGS .......................    33

SIX      THE DONKEY ENGINE ......................    41

SEVEN    THE MENDOCINO RAILROADS ...................    53

EIGHT    THE RAILROADS OF HUMBOLDT ...............    83

NINE     DEL NORTE ...............................   120

TEN      INDEX ...................................   143

# LIST OF ILLUSTRATIONS

*Page*

The redwoods. A contemporary scene in Humboldt County. . . . . . . . . . . .  xvi

Young growth surrounds an ancient stump. . . . . . . . . . . . . . . . . . . . . . . .  2

The quiet serenity of the redwood forest. . . . . . . . . . . . . . . . . . . . . . . . .  3

Scene in Humboldt County in the 1890s. . . . . . . . . . . . . . . . . . . . . . . . . .  3

Testament to the durability of the redwood. . . . . . . . . . . . . . . . . . . . . . .  4

Typical of the publicity shots favored in earlier days. . . . . . . . . . . . . . .  4

Area in the Mad River country that was logged and burned in
   the 1890s . . . . . . . . . . . . . . . . . . . . . . . . . . . . . . . . . . . . . . . . . . . . . . .  5

A diseno (sketch map) of the mid-19th century. . . . . . . . . . . . . . . . . . .  6

Early whipsaw mill. . . . . . . . . . . . . . . . . . . . . . . . . . . . . . . . . . . . . . . . . .  8

Logs had to be raised to this early mill from the skidroad. . . . . . . . . . . .  9

Unusual sawmill, designed to harness the wind. . . . . . . . . . . . . . . . . . .  9

Big water wheel supplied power of early mill. . . . . . . . . . . . . . . . . . . . .  9

Eureka, as painted by a soldier from Fort Humboldt in 1854. . . . . . . . . .  10

Preparing to roll a large redwood log onto an ox-drawn wagon. . . . . . . .  10

Early mill at Eureka. . . . . . . . . . . . . . . . . . . . . . . . . . . . . . . . . . . . . . . . .  11

Logs piled along the banks of Big River. . . . . . . . . . . . . . . . . . . . . . . . .  13

Oxen and horses team up to bring a turn of logs off the hill at
   Big River. . . . . . . . . . . . . . . . . . . . . . . . . . . . . . . . . . . . . . . . . . . . . . . .  14

Horse-drawn tram car at the dump of Stewart, Hunter & Johnson
   on Mill Creek. . . . . . . . . . . . . . . . . . . . . . . . . . . . . . . . . . . . . . . . . . . . .  15

Trinidad, California, in the late 1870s. . . . . . . . . . . . . . . . . . . . . . . . . .  15

Oldest known photo of logging in the redwoods. . . . . . . . . . . . . . . . . . .  16

Early mill at Noyo, just south of Fort Bragg. . . . . . . . . . . . . . . . . . . . . .  17

Early mill and camp in Humboldt County. . . . . . . . . . . . . . . . . . . . . . . .  17

A nester's cabin in the 1880s. . . . . . . . . . . . . . . . . . . . . . . . . . . . . . . . . .  19

Choppers starting on a tree. . . . . . . . . . . . . . . . . . . . . . . . . . . . . . . . . . .  20

It took a lot of chopping to provide the stage for this group. . . . . . . . . . .  20

Choppers at work for the John Vance Mill & Lumber Co. . . . . . . . . . . . .  21

Choppers prepare to start on a schoolmarm. . . . . . . . . . . . . . . . . . . . . . .  21

Men show how a saw is used on occasion. . . . . . . . . . . . . . . . . . . . . . . . .  22

One undercut for three trees. . . . . . . . . . . . . . . . . . . . . . . . . . . . . . . . . . .  22

Choppers prepare to start on a tree high above the ground. . . . . . . . . . . .  22

Choppers and their tools. . . . . . . . . . . . . . . . . . . . . . . . . . . . . . . . . . . . . .  23

The backcut was made with a long saw. . . . . . . . . . . . . . . . . . . . . . . . . . .  23

The fallen tree looked even larger to the buckers. . . . . . . . . . . . . . . . . . .  24

Large logs cut in the operations of John Vance Mill & Lumber
   Company. . . . . . . . . . . . . . . . . . . . . . . . . . . . . . . . . . . . . . . . . . . . . . . . .  24

Page

Fallen snag being bucked for shingle bolts. . . . . . . . . . . . . . . . . . . . . . . . . . 25
Splitting large redwood logs with black powder. . . . . . . . . . . . . . . . . . . . . 25
Logging operations of the Excelsior Redwood Company in
    the 1880s. . . . . . . . . . . . . . . . . . . . . . . . . . . . . . . . . . . . . . . . . . . . . . . . . . 26
Logs stacked in Elk River in 1892. . . . . . . . . . . . . . . . . . . . . . . . . . . . . . . 27
Peelers at work on a log. . . . . . . . . . . . . . . . . . . . . . . . . . . . . . . . . . . . . . . 27
Even after burning, much waste and debris is evident in this scene. . . . 28
Ox team in the Redwood country. . . . . . . . . . . . . . . . . . . . . . . . . . . . . . . . 28
Early horse tram in Humboldt County. . . . . . . . . . . . . . . . . . . . . . . . . . . . 29
A lot of work went into building a skidroad. . . . . . . . . . . . . . . . . . . . . . . 30
A primitive steam locomotive. . . . . . . . . . . . . . . . . . . . . . . . . . . . . . . . . . . 31
Bridging sometimes needed for the bull teams. . . . . . . . . . . . . . . . . . . . . 31
Hand-hewed timbers. . . . . . . . . . . . . . . . . . . . . . . . . . . . . . . . . . . . . . . . . . 32
The loading chute at Nip and Tuck Landing. . . . . . . . . . . . . . . . . . . . . . 33
Loading at the Handly chute at Albion in the 1860s. . . . . . . . . . . . . . . . 34
Point Arena landing, 1882. . . . . . . . . . . . . . . . . . . . . . . . . . . . . . . . . . . . . . 34
Westport Landing was a busy place, despite handicaps. . . . . . . . . . . . . . 35
A busy scene at Albion, dated August 15, 1897. . . . . . . . . . . . . . . . . . . . 35
Bourn's Landing, just north of Gualala, about 1885. . . . . . . . . . . . . . . . . 36
Bourn's Landing about 1903. . . . . . . . . . . . . . . . . . . . . . . . . . . . . . . . . . . . 36
Steamer "Cleone" loading at Mattole Landing. . . . . . . . . . . . . . . . . . . . . 37
The steamer "Brunswick" passing The Heads. . . . . . . . . . . . . . . . . . . . . . 37
Steamer "Caspar" loading at Caspar. . . . . . . . . . . . . . . . . . . . . . . . . . . . . . 37
Steam schooner loading from the wire chute at Caspar. . . . . . . . . . . . . . 38
Robertson raft under construction at Noyo in the early 1890s. . . . . . . . . 39
The landing at Rockport. . . . . . . . . . . . . . . . . . . . . . . . . . . . . . . . . . . . . . . 39
Loading under the wire chute at Noyo. . . . . . . . . . . . . . . . . . . . . . . . . . . . 39
Robertson raft leaving the Noyo River. . . . . . . . . . . . . . . . . . . . . . . . . . . . 40
Westport Landing in 1920. . . . . . . . . . . . . . . . . . . . . . . . . . . . . . . . . . . . . . 40
John Dolbeer. . . . . . . . . . . . . . . . . . . . . . . . . . . . . . . . . . . . . . . . . . . . . . . . . 41
An early Dolbeer donkey. . . . . . . . . . . . . . . . . . . . . . . . . . . . . . . . . . . . . . . 42
Rolling a large log with a Dolbeer donkey. . . . . . . . . . . . . . . . . . . . . . . . 42
Dolbeer used for loading. . . . . . . . . . . . . . . . . . . . . . . . . . . . . . . . . . . . . . . 43
Setting up the donkey to start yarding on a new side. . . . . . . . . . . . . . . 44
Rigging up to yard some oversized logs. . . . . . . . . . . . . . . . . . . . . . . . . . . 44
Later version of the Dolbeer donkey. . . . . . . . . . . . . . . . . . . . . . . . . . . . . 44
Yarding with a Dolbeer. . . . . . . . . . . . . . . . . . . . . . . . . . . . . . . . . . . . . . . . 45
Plenty of blocks and line to provide Dolbeer with leverage. . . . . . . . . . 46
Captain Robert Dollar. . . . . . . . . . . . . . . . . . . . . . . . . . . . . . . . . . . . . . . . . 47
First donkey ever used in the redwoods on a long haul. . . . . . . . . . . . . . 47
Bull donkey for roading logs and Dolbeer for loading. . . . . . . . . . . . . . . 48
Big turn of logs being brought in. . . . . . . . . . . . . . . . . . . . . . . . . . . . . . . . 48
Fine view of an early bull donkey. . . . . . . . . . . . . . . . . . . . . . . . . . . . . . . 49
Patent drawing of the Dolbeer Logging Engine. . . . . . . . . . . . . . . . . . . . 49
Good view of a skidroad with a turn of logs. . . . . . . . . . . . . . . . . . . . . . . 50
Big road engine "California." . . . . . . . . . . . . . . . . . . . . . . . . . . . . . . . . . . . 50
Early logging scene in Humboldt County. . . . . . . . . . . . . . . . . . . . . . . . . 51
Unusual bull donkey. . . . . . . . . . . . . . . . . . . . . . . . . . . . . . . . . . . . . . . . . . 52
"Mrs. Duncan's Teakettle." . . . . . . . . . . . . . . . . . . . . . . . . . . . . . . . . . . . . . 52

*Page*

Glynn & Peterson Lumber Company train at Delmar Landing. ........ 53
Gualala Mill Company No. 2 at the dump. ......................... 54
Gualala Mill Company No. 1. ................................... 54
Gualala Mill Company No. 2 from another angle. ................... 55
The "S. H. Harmon" and the "W. B. Heywood." ................... 55
Gualala Mill Company No. 3. ................................... 56
The Stevenson bridge on the Elk Creek RR. ....................... 57
The Salsig store and post office about 1905. ...................... 58
The Greenwood landing from an early post card. ................... 58
L. E. White Lumber Company No. 1 pauses at the water spout. ........ 59
Later view of the landing at Greenwood. .......................... 59
Albion Lumber Company No. 3. ................................. 60
Mendocino Lumber Company dump scene. ........................ 60
The railroad clung precariously to cliffs above the sea. .............. 61
Mendocino Lumber Company tug boat "Maru." ................... 61
At the L. E. White Lumber Company dump. ....................... 62
Tallying ties at the Rollerville Landing. ........................... 62
Switching lumber cars at the mill near Greenwood. ................. 63
Moving the camp of Navarro Lumber Company. .................... 63
Gualala Mill Company No. 4. ................................... 64
Another view of the mill. ...................................... 64
Mendocino Lumber Company dump in Big River. ................... 65
Unloading locomotive No. 202 of the Northwestern Pacific. ........... 65
View of the mill and harbor at Albion. ........................... 66
Albion River RR. No. 1 with some big logs. ....................... 66
Big mill of the Mendocino Lumber Company. ...................... 67
Loading with the Lawson skyline system. ......................... 67
Caspar Lumber Company at the mouth of Caspar Creek. ............. 68
Caspar Lumber Company dump. ................................ 68
The Mendocino "two spot" with the woods crew. ................... 69
Caspar Lumber Company No. 1, the "Jumbo." ..................... 69
Crossing a trestle on the Albion River RR. ........................ 70
Mendocino Lumber Company No. 1, a steam dummy. ................ 70
Caspar Lumber Company No. 2, "Daisy." ......................... 71
Fort Bragg RR. No. 1, the "Sequoia." ........................... 71
The "Smilax" on the Jug Handle Creek trestle. .................... 72
The "Hercules" in front of the tunnel. ........................... 73
Scene at the head of a Caspar Lumber Company incline. ............. 73
A log train on the Fort Bragg RR., headed by locomotive No. 3. ........ 74
The "Samson" on the Jug Handle Creek trestle in 1925. .............. 74
Early view of the Union Lumber Company plant at Fort Bragg. ........ 75
The "Trojan," first of the Mallets. .............................. 75
Loading in the Caspar woods with the Lawson skyline system. ........ 76
California Western No. 23 at Fort Bragg. .......................... 76
Early operations in the Ten Mile River area. ...................... 76
The Glen Blair Redwood Company mill, as it looked in 1890. ......... 77
Equipment and shops of the Finkbine-Guild Lumber Company
    at Rockport. ............................................... 78
Irvine & Muir used this primitive train at Irmulco. ................. 78

*Page*

California Western train near Camp 10. .......................... 78

Charles Russell Johnson. ........................................ 79

Motor car M-100, one of the famous California Western "Skunks." ..... 79

California Western No. 17. ....................................... 79

The Finkbine-Guild Lumber Company plant at Rockport. ............. 80

Builder's picture of the little Mattole Lumber Company No. 1. ........ 81

Union Lumber Company Shay No. 2. ............................. 81

Comprehensive view of an early railroad landing. .................... 82

The Humboldt Redwoods ........................................ 83

John Vance. .................................................... 83

View of the Big Bonanza mill, taken in the 1880s. ................... 84

Photographer preparing to film company officials. ................... 84

An early landing in the Vance woods on Mad River. ................. 85

Landing scene of the 1890s. ...................................... 85

"Antelope" at the Mad River Slough landing in the 1880s. ............ 86

The mill at West Eureka (Samoa). ................................ 87

The little "Gypsy" works a landing in the woods. ................... 87

"The Astor Cut" section of a redwood tree. ........................ 88

Using the gypsy to load the cars. ................................. 88

Hammond Lumber Company camp in the heyday of railroad logging. .. 89

Andrew Benotti Hammond. ...................................... 89

The Oregon & Eureka RR. No. 11 headed for Samoa. ............... 90

The Jolly Giant mill of Isaac Minor and Noah Falk. ................. 91

This little tank engine was No. 1 on the Arcata & Mad River RR. ...... 91

The Eureka & Klamath River RR. ................................ 92

The Arcata & Mad River RR. No. 3, the "North Fork." .............. 92

The Arcata & Mad River steamer "Alta." .......................... 92

Superintendent V. Zaruba, with his inspection car. .................. 93

The tug "Mary Ann" helps a well-laden schooner. .................. 93

Isaac Minor. ................................................... 94

The "Blue Lake" crossing the North Fork of Mad River. ............. 94

The plant of the Minor Mill & Lumber Company at Glendale. ......... 95

The Arcata & Mad River roundhouse at Arcata about 1890. ........... 95

The "Blue Lake" was a graceful little 2-4-0, shown here at
   Blue Lake. .................................................. 96

Logging railroad served by this little gypsy engine ran north
   into the timber. ............................................. 96

The Arcata & Mad River engine house at Korbel in the early 1900s..... 97

Little locomotive used in the woods by Humboldt Lumber
   Mill Company. .............................................. 97

The logging railroad of Minor Mill & Lumber Company. .............. 98

The Northern Redwood Lumber Company mill at Korbel. ............. 98

Early locomotive built to the Dolbeer design. ...................... 99

No. 11 is bringing a train down from Korbel. ....................... 99

The steamship "Iran" loading at the Arcata wharf in the late 1890s. .... 99

Mill of the Riverside Lumber Company at Korbel about 1890. ......... 100

Two Heislers of Northern Redwood Lumber Company. ............... 100

View of Trinidad in the late 1870s................................ 101

*Page*

The Trinidad Mill Company "Sequoia" at work near Trinidad
    about 1880. . . . . . . . . . . . . . . . . . . . . . . . 101
A Dolbeer locomotive at work near Humboldt Bay. . . . . . . . . . . . . 102
Arcata & Mad River No. 2, as she appeared at Riverside in 1925. . . . . . 102
Loading shingles at Houda's Landing, near Trinidad, in the
    late 1880s. . . . . . . . . . . . . . . . . . . . . . . . 102
David Evans. . . . . . . . . . . . . . . . . . . . . . . . . . . . . 103
The "Emily" at the Trinidad wharf, 1890. . . . . . . . . . . . . . . . 103
Excelsior Redwood Company train from the Freshwater woods. . . . . . . 104
Sailing vessels loading at the Excelsior Redwood Company mill. . . . . . 104
The Humboldt Logging RR. No. 3. . . . . . . . . . . . . . . . . . . . 104
Noah H. Falk. . . . . . . . . . . . . . . . . . . . . . . . . . . . . 105
Three trains pose for the photographer in the Excelsior camp
    at Freshwater. . . . . . . . . . . . . . . . . . . . . . . 105
Unusual articulated locomotive built for the Bayside Lumber
    Company in 1910. . . . . . . . . . . . . . . . . . . . . . 106
Flanigan, Brosnan & Company No. 2 at the dump on Gannon
    Slough. . . . . . . . . . . . . . . . . . . . . . . . . . . 106
Log train of the Bucksport & Elk River RR. at the dump at
    Bucksport. . . . . . . . . . . . . . . . . . . . . . . . . 107
Flanigan, Brosnan & Company trains headed by locomotives
    No. 2 and No. 3. . . . . . . . . . . . . . . . . . . . . . 107
Along Ryan Creek at the eastern edge of Eureka. . . . . . . . . . . . . 108
Early rod engine built by Globe of San Francisco. . . . . . . . . . . . . 108
The Elk River mill at Falk in 1888. . . . . . . . . . . . . . . . . . . 109
The sawmill of Elk River Mill & Lumber Company. . . . . . . . . . . . 109
No. 2 of the Elk River Mill & Lumber Company at Falk. . . . . . . . . 109
Elk River Mill & Lumber Company No. 1, the "Falk." . . . . . . . . . . 109
Dolbeer & Carson No. 2 with a train. . . . . . . . . . . . . . . . . . 110
Baldwin locomotive "Sequoia." . . . . . . . . . . . . . . . . . . . . . 110
McKay & Company's 1 Spot was turned out by Globe of
    San Francisco. . . . . . . . . . . . . . . . . . . . . . . 110
Loggers were an inventive lot and were constantly experimenting. . . . . . 111
Builder's photo of McKay & Company's No. 2. . . . . . . . . . . . . . 111
William Carson. . . . . . . . . . . . . . . . . . . . . . . . . . . . 112
Dolbeer & Carson No. 3 with a load of redwood logs. . . . . . . . . . . 112
The world-famous Carson mansion in Eureka. . . . . . . . . . . . . . 113
Peter J. Rutledge. . . . . . . . . . . . . . . . . . . . . . . . . . . 114
San Francisco & Northwestern No. 5, at work at Happy Camp
    near Holmes. . . . . . . . . . . . . . . . . . . . . . . . 114
Simon Jones Murphy. . . . . . . . . . . . . . . . . . . . . . . . . . 115
Pacific Lumber Company No. 1 on a passenger train at Scotia. . . . . . . 115
Early picture of Pacific Lumber Company No. 2. . . . . . . . . . . . . . 116
San Francisco & Northwestern RR. No. 0, called "The Dummy." . . . . . 116
View of Scotia and the big mill of The Pacific Lumber Company
    about 1902. . . . . . . . . . . . . . . . . . . . . . . . . 116
Pacific Lumber Company train on Stitz Creek trestle in 1895. . . . . . . 117
Pacific Lumber Company No. 21, a classic Rogers 4-4-0. . . . . . . . . . 117
Pacific Lumber Company No. 1 is dwarfed by a big log. . . . . . . . . . 117
The little Excelsior Redwood Company "Deuce" now sports a tender. . 118

*Page*

Pacific Lumber Company train in the woods. ....................... 118

Locomotive No. 25 trundles a train across the Holmes-Shively
    bridge. ...................................................... 118

Pile driver at work on a trestle for the Pacific Lumber Company. ...... 119

Albert Stanwood Murphy. ......................................... 119

Camp of The Pacific Lumber Company in the Freshwater operations. ... 119

A Hobbs, Wall & Company Shay poses high atop a bridge of
    cribbed logs. ................................................. 120

Little Forney locomotive handles a log train for Hobbs, Wall &
    Company. .................................................... 121

Hobbs, Wall & Company train near Crescent City. ................... 121

Camp buildings of the Brookings Timber & Lumber Company. ........ 121

Little Hobbs, Wall engine switching the yards at Crescent City. ....... 122

The California & Oregon Lumber Company poses the No. 5 on
    the Chetco River Bridge. ...................................... 123

California & Oregon Lumber Company train headed for
    Brookings, Oregon. ........................................... 124

The big mill at Brookings, as seen from the ocean side. ............. 125

Wharf scene at Brookings, Oregon. ................................ 125

Wharf of the California & Oregon Lumber Company at Brookings,
    Oregon. ..................................................... 126

The modern day chopper uses a gasoline-powered chain saw. ......... 127

Howard A. Libbey. ............................................... 128

Sighting the fall with a gunstick. ................................. 128

With the undercut finished, the chopper starts the backcut. ........... 129

Timmmberrr! As the big tree starts its downward course, choppers
    scramble to safety. ........................................... 130

On the ground at last, the big tree dwarfs the choppers. ............. 131

Setting chokers on a cat show. .................................... 131

Yarding with a cat and arch at Big Lagoon, August 1956. ............. 132

Big Lagoon, August 1956. ......................................... 133

The cats displaced the donkeys in the 1930s. ....................... 133

Stanwood A. Murphy. ............................................. 134

Probably the heaviest loader ever built. ............................ 134

The old and the new at The Pacific Lumber Company. ................ 135

A. F. "Bud" Peterson. ............................................ 136

Crossing the Eel River on the private truck road of the Pacific
    Lumber Company. ............................................ 136

Loading some big ones with a crotch line at Simpson Timber Company. .. 137

Hammond logs travel the coast highway near Big Lagoon. ............. 137

Topping a spar tree in the 1930s. ................................. 138

Hammond truck landing at Big Lagoon. ............................ 139

Tightening up the binders before heading out on the highway. ........ 140

Modern day loading with a cat and grapple. ........................ 141

Aerial view of Humboldt Bay, looking north. ........................ 142

# ACKNOWLEDGMENTS

ALTHOUGH THE CALIFORNIA redwood lumber industry is one of the oldest industries in the state, it has been largely ignored by the state's historians. The redwood industry has been an important contributor to the economy of California for almost one hundred and twenty-five years. The uniqueness of its product and its location make its history a fascinating story. A historical-pictorial book of redwood logging could not be created alone. The authors would like to gratefully acknowledge the assistance of the following: George B. Abdill, Lois Bishop, Grace Brambani, Walter Casler, Mrs. Dick Childs, The Clarke Memorial Museum in Eureka, E. D. Culp, John Cummings, Edward Freitas, James Gertz, A. W. Giffillan, Gus Haagmark, Lester Holmes, The Humboldt County Historical Society, Michael Koch, Robert J. Lee, John E. Lewis, Gordon Manary, The Oregon Historical Society, James Palmer, Gus Peterson, Martha Roscoe, Jean Shearman, Jack Slattery, Henry Sorenson, Norton Steenfott, Dave Swanlund, Sam Swanlund, and Amos Tinkey.

The authors are also indebted to the following corporate entities: the Arcata Redwood Company, the California Redwood Association, the Georgia-Pacific Corporation, The Pacific Lumber Company, and the Simpson Timber Company.

Augustus William Ericson, the photographer of the early Redwood Country, recorded in picture the early lumber industry, and only recently has he received recognition for his excellent photographs. His daughter, Mrs. Ella Ericson Bryan, helped in identifying many of his photographs. Special thanks should also go to the following: Lockwood Dennis who drew the picture of the whipsaw mill; Mrs. Nannie Escola who graciously offered her photo library on early Mendocino; D. S. Richter and B. H. Ward who helped to unravel ownerships and locomotive rosters; Richard H. Tooker who contributed knowledge of the lumber landings along the coast and the schooners that served them; Gerald M. Best who provided help with the locomotive histories; Harold Mentzer who helped with the history of Mendocino County; and Kramer Adams who assisted with the early history of lumbering.

We are also indebted to Frances Purser of the Humboldt State University Library who collected the A. W. Ericson photographs, Mrs. Marlys Maher of the College of the Redwoods Library for her assistance in supplying materials, and the Bancroft Library of the University of California at Berkeley.

Lynwood Carranco
John T. Labbe

# LOGGING THE REDWOODS

*The redwoods. A contemporary scene in Humboldt County, little changed in the last few million years.*

# CHAPTER ONE

# THE REDWOODS

SEQUOIA SEMPERVIRENS, the coast redwood of California, dates far back into the ages of the past. More than one hundred million years ago a vast redwood forest flourished throughout much of the northern hemisphere. Fossil remains of the trees have been unearthed in Alaska, Greenland, and Spitzbergen, as well as most of the United States and Europe. During the millenia that followed, dramatic changes took place in the land formations and climate, but the Sequoia managed to hold on and survive. Rather than adapt to the changing conditions, these trees preferred to retreat to areas where the growing conditions were more to their liking, until today they are confined to the northern coast of California. Other scattered remnants of the Sequoia family are still to be found in China, Japan and Formosa.

The sempervirens require a temperate climate, rich soil, and plenty of moisture. And here, where the westerlies come in moisture-laden from the Pacific to meet the heated air of the interior valleys, heavy fogs provide a large measure of the water they require. The trees are found within this fog belt from south of Monterey Bay to the Oregon border, reaching inland no more than thirty miles at points where the fog penetrates. A few scattered groves extend south as far as the northern part of San Luis Obispo County, while Oregon lays claim to those that reach north to the Chetco River.

Here, within the confines of this fog belt extending roughly five hundred miles in length, the great trees enjoy a healthy growth. The coast redwood is probably the most vigorous of all the commercial trees in the West. It is extremely resistant to such natural enemies as fire, insects, and disease, and normally reaches maturity in from four hundred to eight hundred years. But perhaps its greatest asset lies in its ability to reproduce from sprouts. Most of the young trees receive their start in this manner, while only a small percentage reproduce from seed. Thus, new trees are able to take advantage of a root system that may have nourished several generations of trees.

Under ideal growing conditions young trees may add as much as an inch in diameter and two feet in height each year. A three-foot redwood tree may be but thirty-six years old, while another of the same diameter, under less favorable growing conditions, might be well over three hundred years of age.

The sempervirens is a spectacular tree by any standards. It is the tallest of all trees, reaching a height of three hundred fifty feet. The tallest known specimen is the H. A. Libbey tree, which has been measured at three hundred seventy-two feet. In diameter, they may reach twelve to sixteen feet, although the average is but four or five feet. The largest measured is the Stout Tree in Jedediah Smith State Park with a diameter of twenty feet. It is not unusual for a tree to reach a great age where it is protected from such natural disasters as wind and floods. The oldest tree so far measured had a life span of two thousand two hundred years.

The resistance of the redwood to the ravages of time is perhaps its most outstanding attribute. Loggers have found sound logs that have lain on the forest floor for so long that mature trees have grown over them, yet they have produced good lumber. In fact, many of the logs that the early loggers had to abandon in the cruder operations of a century ago have been salvaged by the modern logger, and have provided top quality lumber. These characteristics carry over into the finished lumber and help make redwood one of the most valuable of all wood products.

California is also host to another unique

*A. W. Ericson photo*

*Young growth surrounds an ancient stump.*

*To stand in the quiet serenity of the redwood forest is a spiritual experience, akin to standing in the nave of a great cathedral.*

*Scene in Humboldt County in the 1890s. Note the old stumps, evidence of earlier logging that has made but little impression on the forest.*

redwood tree. This tree is found along the western slopes of the Sierra Nevada, at an elevation of five thousand to eight thousand feet, in an area extending from Placer County in the north to Tulare County in the south. Known as the Sequoiadendron giganteum, it appears to the casual observer to be quite similar to the Sequoia sempervirens, but genetically they are very different. Like the sempervirens, the giganteum is very resistant to fire and insect damage, and it has the same reddish coloring.

The giganteum is a shorter tree, the tallest being measured at three hundred ten feet, but it reaches a far greater diameter at the base, which gives it a more tapered bole. The General Sherman Tree is con-

*Courtesy the Pacific Lumber Company*
*Testament to the durability of the redwood, despite the questionable age attributed to the stump.*

*Clarke Memorial Museum*
*Typical of the publicity shots favored in earlier days to impress friends in the East with the size of western trees.*

sidered to be the most massive tree on earth. It has a diameter at the base of 32.2 feet, and is probably three thousand six hundred years old. These trees normally reach a greater age than the sempervirens, and they grow more slowly. Several are known to be more than three thousand years old. But perhaps the greatest difference lies in the fact that the giganteum reproduces only from seed, which makes its future on this planet far more tenuous.

A. W. Ericson photo

*Area in the Mad River country that was logged and burned in the 1890s. The regenerative power of the trees is evident in the new sprouts springing from the stumps.*

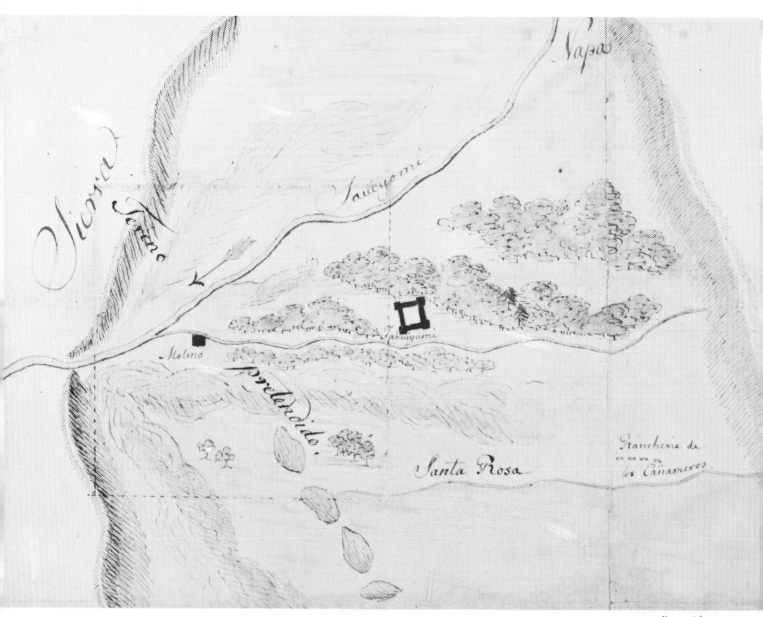

*A diseño (sketch map) of the mid-19th century shows a portion of Cooper's Molino Rancho at a time when the Mexican grant was undergoing court test. The Russian River is here identified as Tauiyomi, while the fort-like compound built in the 1830s by colonists from Mission Sonoma on Mark West Creek, is labeled Tahuiyumi. (Bancroft Library.)*

# CHAPTER TWO

# EARLY SAWMILLS

THE FIRST RECORDED sighting of the redwood trees was noted in 1769. In that year the Spaniards sent the Portola Expedition overland from San Diego in a search for Monterey Bay. The expedition arrived in due course, but failing to recognize the bay, continued north along the coast. Here, in the vicinity of Santa Cruz, they came upon these strange new trees, and this was duly noted in the diary of Father Crespi on October 10th. The Spaniards called them *palo colorado*, or red wood.

Within the next two decades the Spaniards pushed their settlements north to the San Francisco area. In their buildings they continued to favor their traditional architecture, using adobe walls and tile roofs. However, a certain amount of wood was required for timbers and beams. Probably the first use of redwood for construction purposes was in the establishment of the Mission Santa Clara, as well as Mission Dolores and the Presidio of San Francisco. The timbers were no doubt hewn or split for the purpose, probably by Indian labor.

In 1812 the Russians, working south from their base in Alaska, established a foothold on the coast north of San Francisco Bay. Here they built a settlement at Fort Ross. Unlike the Spaniards, the Russians were accustomed to build with wood, and most of their structures were constructed of lumber. They used the redwood growing on the site. The lumber was split and sawed, while shingles were split for roofing. Their sawmills consisted of a pit, over which a framework was erected to support the log. Two men sawed the boards by hand, one standing atop the log while the other worked in the pit below. These pitsaw or whipsaw mills were common throughout the world at that time, and are still to be found in some primitive areas today.

In addition to cutting for the needs of the new colony, some lumber was probably shipped to outposts in the north. Several ships were constructed of redwood at Fort Ross, but the experience proved far from satisfactory, largely because of the failure of the Russians to properly season the wood before using it in the ships.

Meanwhile, the Spanish developments along the peninsula south of San Francisco, as well as at the Presidio, received supplies of timber from cuttings established in the forest around Woodside, to the west of Palo Alto. In this area along San Francisquito Creek, Indian labor provided for the needs of the missions, but transportation to the Presidio was difficult, as the timbers had to be dragged or carried over primitive roads, and by 1820 most of the supply came from Marin, on the north side of the bay.

By about 1830 the demand for lumber had increased to the point where it became attractive to individuals seeking a means of livelihood. For the most part, these were men who had deserted from American and British ships. They had sought refuge in the forests where game was plentiful and living was comparatively easy. To provide a small income they began sawing lumber. This was a profitable occupation that required a minimum of their time. Lumber was bringing fifty dollars a thousand board feet, and two men could whipsaw one hundred feet a day. Shingles were worth seven dollars a thousand. With very little labor these early lumbermen were able to provide for their needs. The trees were

there for the taking, and their only problem seems to have been the marauding bears that raided their supplies and became trapped in the sawpits.

The Monterey and Santa Cruz districts, already established as cutting areas, were the sites of the first whipsaw mills, but others quickly sprang up in the East Bay, Marin County, and Santa Clara County. Transportation was a greater problem around Santa Cruz, and most of the early cut was to supply the needs of the area. As the demand for lumber at San Francisco increased, it was found profitable to ship some of the output north by sea.

The first waterpower mill appears to

*L. Dennis*

*Early whipsaw mill of the type that dominated the scene during the days before the gold rush, when Spain ruled California.*

*Logs had to be raised to this early mill from the skidroad. Location of the mill was probably dictated by the available water supply.*

*This unusual sawmill, designed to harness the wind for motive power, bears witness to the ingenuity of the early lumbermen. Built at Mendocino in the early 1870s by the local banker, it proved a failure.*

*Dominant feature of this early mill is the big water wheel that supplied the power.*

*Humboldt County Historical Society*

*Eureka, as painted by a soldier from Fort Humboldt in 1854. Prominent on the shore is the Ryan, Duff & Company sawmill. Smaller mills are shown on the shore to the left.*

*Freese & Fetrow photo from M. Koch*

*Preparing to roll a large redwood log onto an ox-drawn wagon. This was an unusual type of transportation in the redwood region, where the rough ground and constant moisture would have made the going difficult for such a vehicle. Presumably, it was only used to log odd tracts in and around Eureka.*

have been used at Mission San Gabriel, in the Los Angeles area far from the red-woods, in 1823. Powered mills were slow in coming to the San Francisco district. The early settlers were not lumbermen, and their needs were small. The demand was easily met by the whipsawyers, or by Indian labor, which was plentiful and cheap.

It was 1834 before the first commercial waterpower mill was put in operation by John B. R. Cooper. This was a combination grist and sawmill that was built on the south bank of Mark West Creek, once called Mill Creek, about one thousand feet from its confluence with the Russian River. This outpost in Sonoma County, some distance from the populated areas, may have been designed as a barrier against Russian expansion. But, however it came about, it represented an investment of ten thousand dollars, a sizable sum for the time. During the winter of 1840-1841, the mill was washed away by high water and was not rebuilt.

About this same time, another water-power mill was erected near Mt. Hermon, in Santa Cruz County, by the Danish blacksmith Peter Lassen. Lassen never operated the mill himself, choosing instead to trade it off to Isaac Graham for one hundred mules. Graham put the mill in operation in 1842 and continued to operate it for a number of years. Lumber was now bringing up to two hundred dollars a thousand feet at San Francisco, and a large part of his production was shipped there.

*Early mill at Eureka. On the left, the ox-drawn wagon has just unloaded beside the horse-drawn tram car on wooden rails that handles the regular log supply to the mill.*

These waterpower mills employed a vertical saw that cut in much the same manner as the whipsaws, except that a waterwheel provided the power to work the saw blade up and down. These were known as "muley" mills. With his muley mill Isaac Graham was able to turn out about five hundred feet of lumber a day on the Zayante Grant. It was the opening wedge in a series of small operations in the San Lorenzo River drainage, one of the finest stands of virgin timber in the West. And in this connection it is interesting to note that lumbering was only recently brought to a close in this same area, after nearly one hundred forty years. Moreover, it was not brought about by the lack of available timber, but rather by demands for development and recreation that made continuing operations both unprofitable and unpopular.

In 1843 Stephen Smith brought about a revolution in redwood lumbering when he arrived at Bodega, in Sonoma County, with the first steam engine in California. He set up a saw and grist mill at Bodega that went into operation in 1844. The mill operated for ten years, when it was destroyed by fire. It was not rebuilt.

During the 1840s, the production of lumber was stimulated by a rapidly growing demand. Whipsaw mills continued to dominate the scene, with an occasional waterpower mill springing up, spurred on by the high prices offered in the San Francisco area. In the Redwood Canyon area east of Oakland, where transportation was less difficult than along the peninsula, new mills began to blossom. The timber supply was somewhat limited, but the growing demand and high prices soon made this the major industry of the East Bay area.

# CHAPTER THREE

# OPENING THE NORTH COAST

THE DISCOVERY of gold in 1849 brought sudden and dramatic changes in the lumber industry. The admittance of California to the Union in 1850 placed the timber resources of the region for the first time under restrictive government regulation. But under the impetus of the gold fever, it was some time before the new government could make its influence felt, thereby minimizing the impact of the change.

With the first cry of "gold," the sawyers abandoned their sawpits and flocked to the diggings. At the same time, immi-grants began to converge on the state in growing numbers. The demand for lumber soared, as did the prices. Almost overnight the promise of fortunes in lumbering attracted a new breed of men, men with capital and experience who would take full advantage of the opportunity.

The new mills were commercial enterprises designed to supply the expanding markets. Many of them were run by waterpower, although the use of steam was increasing. All of the mills, to begin with, were muley mills with a sash saw or gang of saws.

Operations began in the area around

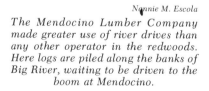

*Nannie M. Escola*

*The Mendocino Lumber Company made greater use of river drives than any other operator in the redwoods. Here logs are piled along the banks of Big River, waiting to be driven to the boom at Mendocino.*

*Oxen and horses team up to bring a turn of logs off the hill at Big River, near Mendocino. This was Camp A of the Mendocino Lumber Company in 1902.*

San Francisco, but soon moved up the coast, where great stands of timber grew close to the sea. A steam mill was built in Contra Costa County in 1849, followed shortly by others. In the same year, Judge Blackburn built a mill near Santa Cruz, but the rush to the goldfields forced the prices of materials and labor to dizzying heights, and the enterprise proved a failure before a wheel was turned.

In 1850, the Gregg expedition, seeking a shorter supply route to the Trinity County mining areas, stumbled on Humboldt Bay. Until this discovery, a more dubious anchorage in the lee of Trinidad Head had offered the greatest promise. Settlement of the Humboldt Bay area followed very quickly. The Pioneer mill was built by Eddy & White during the summer of 1850, although it did not prove to be particularly successful. This was followed immediately by a number of others.

The first really successful mill on the bay was that of Ryan, Duff & Company, which went into operation in 1852. These men had purchased the old sidewheeler SANTA CLARA in San Francisco and had

*Nannie M. Escola*

*Horse-drawn tram car at the dump of Stewart, Hunter & Johnson on Mill Creek, a tributary of Ten Mile River, north of Fort Bragg, in 1883. It was in this year that the young C. R. Johnson joined Stewart & Hunter. From this beginning emerged the great Union Lumber Company operation at Fort Bragg.*

*A. W. Ericson photo*

*Trinidad, California, in the late 1870s. To the left is a spindly trestle that carried the railroad across the river.*

loaded her with machinery for a mill. A slough was cut inland from the bay where the city of Eureka now stands, and the steamer was run in and grounded. The sawmill was built alongside and was powered by the engines of the SANTA CLARA through a connection with the main shaft. The ship itself was used as living quarters for the men. It was equipped with four gang saws and produced sixty thousand feet of lumber and forty thousand laths daily. A crew of forty men was employed. As with so many mills, it was soon destroyed by fire, but was quickly rebuilt on a site nearby. By 1854 there were no less than nine mills operating on Humboldt Bay.

Meantime, lumbering along the Mendocino coast was equally active. Despite the lack of formal harbors, the lumbermen moved in their machinery and set to work. Learning of the timber stands near Big River, a group hiked overland from Bodega to see for themselves. They were impressed, and in July of 1852 they returned with machinery for a mill. This was erected on a bluff above Big River. Logs, which were floated down the river,

*Oldest known photo of logging in the redwoods, taken on the Albion River in 1853. Log was being rolled out of the woods by oxen.*

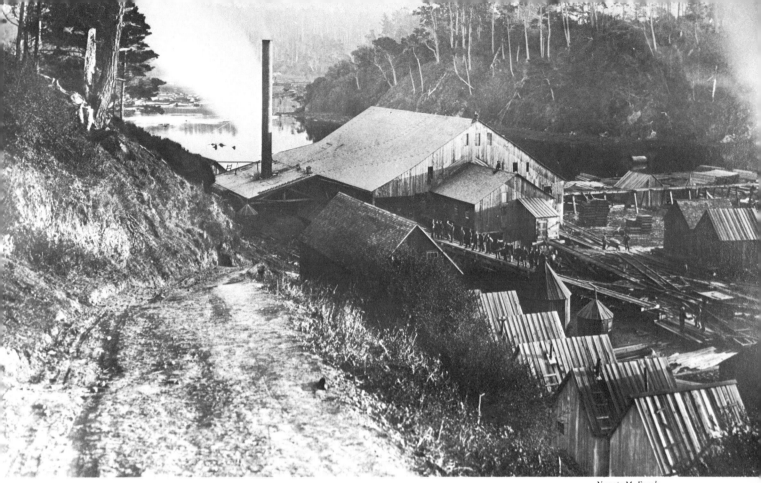

*Early mill at Noyo, just south of Fort Bragg.*

*Early mill and camp in Humboldt County, surrounded by decked logs.*

had to be hoisted up an incline to the mill. The mill proved to be too small for the big redwood logs, and as soon as possible a larger mill was obtained from the East. This new mill was located on the north bank of the river a half mile above the present town of Mendocino.

The first mill had a sash gang with twenty-eight saws. The second mill had two circular saws, a muley, and a sash, and it had a capacity of forty thousand feet of lumber a day. Both mills were in operation until 1856, when the upper mill was removed. This operation, which ultimately became the Mendocino Lumber Company, operated for many years.

A year after the first mill was built at Mendocino, a mill was built at Albion by Dallas, Davidson & McPherson. This was a small steam mill, but it foreshadowed the Albion Lumber Company operation, which became one of the major operations of the area.

All of these early mills were of the muley type, but the erection of the first sawmill using circular saws in 1854 heralded a new era for the lumbermen. This mill was erected near Woodside, the site of some of the original redwood cuttings. The circle mill at Mendocino followed closely, and the muley mills, like the old whipsaw mills, soon disappeared.

# CHAPTER FOUR

# EARLY LOGGING

THE FIRST MEN to challenge the redwoods were ill-equipped for the task. Such massive trees had never before been encountered, and no tools had ever been developed for the purpose. Armed with little more than an ax, the choppers set about their job. It was not unusual for two men to spend as much as a week of steady chopping to bring down one of these giants.

Once on the ground, many more days of work were required to break the logs down into chunks that could be handled. Due to their great weight, the trees usually shattered on impact with the ground. Despite the waste, the broken trees were more easily handled. If perchance a tree failed to break, it was necessary to go up the log to the point where a saw could reach through. This might be as much as two hundred feet from the stump. And once the top had been bucked off, the choppers set to work with hardwood wedges and iron-bound wooden mallets. They split the great log from end to end,

A. W. Ericson photo

*A nester's cabin in the 1880s. These men established claims to the timber lands, later selling them to the lumbermen or to speculators.*

*Choppers starting on a tree. The regular choppers are Hugh McCormick and Bill Chaffey. Posing in the center with an ax is timekeeper Isaac Cullberg, who died in 1961.*

*It took a lot of chopping to provide the stage for this group.*

A. W. Ericson photo

*Choppers at work for the John Vance Mill & Lumber Co. near Arcata in the 1880s.*

A. W. Ericson photo

*With the staging in place, choppers prepare to start on a schoolmarm. From all appearances, the tree on the left is only a snag.*

*Men show how a saw is used on occasion. Large notches have been chopped into the sides of the tree to let the saw clear. The bevel cut into the stump is for the purpose of throwing the tree clear of the stump as it falls, thus reducing the danger of breakage.*

*One undercut for three trees. A tricky situation, as the trees could fall separately, splitting apart and giving the choppers some anxious moments.*

*Choppers prepare to start on a tree high above the ground. Several tiers of staging supported by driving boards raise the choppers to the level of their work. The reason for the high stump is not readily apparent. It may have been to clear the swell of the butt, or the stump may have contained some rot or other defect. The driving boards were commonly called springboards by the Northwest loggers.*

reducing it to sections that could be sawed to length.

The chunks that had been bucked off had to be refined still further before they could be sawn into boards. They were split and hewed into cants, and these cants were worked into place on the scaffolding over the sawpits. Finally the sawyers were ready to reduce the cants to boards and planks with their whipsaws.

With the coming of more sophisticated commercial mills, new logging methods had to be devised to keep them supplied. The larger mills implied a certain permanence, with the result that logs had to be transported from greater distances.

A. W. Ericson photo

*Choppers and their tools. The "gun stick" behind the men lying in the cut was used to sight the direction of fall.*

A. W. Ericson photo

*After the undercut had been chopped, the backcut was made with the long saw held by the man on the stump.*

A. W. Ericson photo

*If the standing tree looked large to the choppers, the fallen tree looked even larger to the buckers.*

A. W. Ericson photo

*Large logs cut in the operations of John Vance Mill & Lumber Company on Mad River. The logs were thirteen feet to nineteen feet in diameter, and the tree was approximately three hundred feet long.*

A. W. Ericson photo

*Fallen snag being bucked for shingle bolts.*

Sam Swanlund

*Splitting large redwood logs with black powder to reduce them to a more manageable size.*

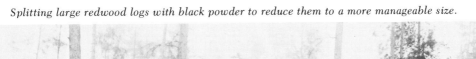

Teams of oxen supplied the power, and at first the logs were rolled to their destination. Later, as the timber receded from the mill site, skid roads had to be built.

The reduction of waste now became a factor. It was easier to transport the logs when they remained unbroken. It was also economically attractive to obtain as much material as possible from each setting. As a result, steps were taken to reduce breakage, and the logger soon learned the value of planning the fall of each tree carefully. Every effort was made to choose the best area in which to lay the tree, and a bed of limbs and underbrush was prepared to cushion the fall.

Among the first requirements were longer and heavier saws. But it was not until the close of the Civil War, out of which came advances in metallurgy, that adequate saws could be designed.

The bark of the redwood is very tough and stringy, and is almost impossible to

A. W. Ericson photo

*Logging operations of the Excelsior Redwood Company in the 1880s. This was in the Freshwater area, just east of Eureka. After the logs were bucked and peeled, the area was burned before the logs were yarded out.*

*Logs stacked in Elk River in 1892, waiting for a winter freshet to carry them down to the booms.*

*Peelers at work on a log.*

*Even after burning, much waste and debris is evident in this scene.*

*Some loggers in the Redwood Country were slow to give up their ox teams. Up until 1915 they were still to be found on some skidroads.*

saw. It not only impeded the cutting action of the saws but had a tendency to become entangled in the machinery. It was necessary, therefore, to peel the bark from the logs before they could be reduced in the mill, a requirement that was unique with the redwood mills. In the modern sawmills of the 1970s, the removal and disposition of the bark is still a major problem.

Since the teams of oxen had to drag the logs long distances along the skidroads, it was the usual practice to peel the logs in the woods. The smooth logs were more easily transported, and the bark and waste were left in the woods.

This accumulation of waste tended to impede logging operations, and the redwood loggers developed a method of dealing with the problem not used anywhere else. When the logs were ready for yarding, the loggers set fire to the waste

and burned the logging area as cleanly as possible. This system was practical, primarily because of the resistance of the wood to fire. And owing to the continual dampness, there was little danger of the fire spreading out of control. While it probably received little or no consideration, the reproductive cycle of the trees remained unaffected. Almost before the ashes had cooled, the stumps began to send out new shoots that would soon become new trees.

Most of the mills were established on streams or rivers where water was available for the mills and millponds. The time-honored method of moving logs in New England, from where most of the early lumbermen came, was by water. The logs were worked into the stream and driven to the mills, often with the aid of splash dams. Water was stored behind the dams and suddenly released to wash the

*Freese & Fetrow photo*
*From M. Koch*

*Early horse team of the Eel River Lumber Company. The mill was at Newburg, east of Fortuna.*

accumulated logs down to the booms. This method was used to some extent in the redwoods, but the results were far from satisfactory. Few of the streams were large enough for drives, and freshets were unpredictable due to the lack of a winter snow pack. In addition, the redwood stores up great quantities of water which makes it extremely heavy. Many of the logs, particularly the butt logs, are too heavy to float. This, together with their size, makes driving impractical. As a result, the loggers turned to tram roads as an extension of their skidroads.

Many of the tram roads were constructed of poles laid in place of rails, and cars with large concave wheels designed to fit over the poles were hauled by horses or oxen. Some of the tram roads, however, were constructed with regulation steel rails in the standard railroad tradition. Around Humboldt Bay, where the land was flat and marshy for some distance between the timber and the bay, tram roads were a necessity. By the fall of 1854, only some four years after the first mill was built, *The Humboldt Times* reported that there were "upwards of 20 miles of good and substantial roads" conveying logs to the water's edge.

*Hammond Lumber Company*

*A lot of work went into building a skidroad. After the logs were placed, the area between the logs had to be filled in to provide footing for the bulls. Costs ran as much as five thousand dollars a mile.*

*Freese & Fetrow photo*
*From M. Koch*

On this early Humboldt County pole tram some enterprising logger has replaced his horse teams with a
primitive steam locomotive.

*A. W. Ericson photo*

Adding to the high cost of skidroads was the extensive bridging sometimes needed to maintain a reasonable grade
for the bull teams.

*Many of the old-timers were artists with the ax and "shin hoe," as demonstrated here in these hand-hewed timbers.*

# CHAPTER FIVE

# MENDOCINO LANDINGS

BECAUSE OF ITS rough and rocky nature, land travel along the Mendocino coast was very difficult. It was therefore necessary for each mill operator to develop some means of getting his product aboard a ship in the vicinity of his mill. Despite the lack of formal harbors, ships were loaded all along the coast. Nearly sixty landings have been identified between Bodega Head and the Humboldt Bar.

Loading facilities required only a secure footing on some rocky outcrop for a chute, where lumber and cargo could be transferred to a small schooner riding just offshore. These loading points were sometimes referred to as "dog-holes," a derogatory term of the day popularly applied to any unattractive location.

A chute was a simple affair, consisting of the wooden chute and the supporting structure. The outer end of the chute could be raised or lowered to suit conditions. Lumber was sent down the chute to crewmen waiting on the ship, who stowed it by hand. A brake, consisting of a hinged gate shod with iron, rode within the chute. The lumber slid under the brake, which slowed its progress before it reached the deck. A rope attached to the brake could be used to raise it to lessen the resistance. At the bottom of the chute a clapper was provided. This was an added brake that was operated by the clapperman, and its function was to stop the boards where the men on the ship could conveniently take hold of them. Usually the clapperman controlled the brake as well.

The chutes were operated in both directions. Incoming cargo went up them in a sled that rode within the chute. Passengers usually boarded by means of small boats.

To bring a schooner into one of these

*The loading chute at Nip and Tuck Landing, a few miles south of Point Arena. This view gives a close look at a typical lumber chute.*

*Loading at the Handly chute at Albion in the 1860s. Lumber and ties are stored in the yard on the bluff.*

*Point Arena landing, 1882. Schooners wait just beyond the line of surf.*

landings was not quite as difficult as it might at first appear. Buoys made of floating logs were anchored offshore, and as the ships arrived, the crew slipped a mooring line through a hole in the buoy and made it fast. Other lines were then carried ashore by men in a small boat, where they were made fast. The ship could then be warped into position and secured.

Since this coastline has always been subject to unpredictable weather, there was the constant danger from sudden storms. If a storm blew up while a ship was moored offshore, bow and stern anchors were dropped and the crew made

*Nannie M. Escola*

*Westport Landing was a busy place, despite the obvious handicaps.*

*Nannie M. Escola*

*A busy scene at Albion, dated August 15, 1897. The ships are taking on varying loads. The one on the left has a load of tanbark, as does the one nearest the wharf. Second from the left is a load of railroad ties, while the third ship is loading lumber.*

for shore. Usually, the chain for the bow anchor was taken around the mast and shackled for additional holding power. If a ship happened to be loading under a chute, every effort was made to warp her back out to the moorage, in order to avoid possible damage to the landing. Once the schooner was secure, the crew made for shore and left her to ride it out alone.

In spite of the inhospitable shores and precarious loading facilities, the schooners provided the necessary link between the Mendocino coast and civilization. Once begun, the lumber industry flourished and kept the little ships scurrying back and forth. In the period between

*Nannie M. Escola*

*Bourn's Landing, just north of Gualala, about 1885. The little schooners had to nestle up against the rocks to load.*

*Robert J. Lee print*
*Nannie M. Escola*

*The wire chute at Bourn's Landing about 1903.*

*Nannie M. Escola*

*Steamer "Cleone" loading at Mattole Landing. Because the landing was so low, it put a severe strain on the ship's rigging, which made it unpopular with many skippers. A higher landing put a more direct pull on the offshore anchor and less strain on the ship's boom.*

*Nannie M. Escola*

*The steamer "Brunswick" passing The Heads, as she steams out of the harbor at Ft. Bragg.*

*Nannie M. Escola*

*Steamer "Caspar" loading at Caspar. Passengers were often transferred on platforms carried in the same manner as the lumber pictured here. For this reason, the wire chutes were often referred to as trapeze chutes.*

1860 and 1864, for instance, some three hundred schooners were in this service. Each ship carried a crew of five or six men and was capable of transporting from ninety to a hundred thousand feet of lumber on each trip.

With the coming of steam in the 1880s the picture changed somewhat. When the first steam schooners appeared early in the decade, they provided larger and more reliable transport. With their steam-powered engines they were better able to maneuver in the cramped moorages, and in case of a storm could cut and run. In addition to their larger cargoes, they provided passengers with more comfortable accommodations.

Steam also brought changes to the landings. Wire chutes began to take the place of the wooden chutes, allowing ships to load farther offshore and in harbors not adaptable to the shore-based slides. The wire chutes consisted of a heavy cable stretched between the ship and shore, over which a carrier traveled with the cargo to be loaded. In design and operation it bore a close resemblance to the skylines used by the loggers in the woods.

In rigging a wire chute, a heavy offshore line was anchored some distance beyond the moorage, and the end of the line was attached to a buoy. Once properly moored, the ship could pick up the offshore line, while the inshore line was brought out from the landing. Usually a light line was first carried out by a small boat, and this was used to bring out the heavier inshore line. The offshore and inshore lines were fastened together on deck with a sliphook, after which the assembly was raised into position through a block attached to a boom on the ship. Thus, the main stress was against the offshore anchor while the carrier traveled back and forth between ship and shore with the cargo. Most wire chutes were operated by steam-powered drums at the shore end, although some were operated by counterweights. In case of a sudden change in the weather, the steamers moved to safer positions offshore. A sharp blow to the sliphook dropped the carrier lines, which saved much valuable time.

Gasoline-powered schooners appeared in small numbers in the 1890s. Being smaller and less powerful, they were usu-

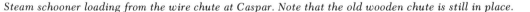

*Nannie M. Escola*

*Steam schooner loading from the wire chute at Caspar. Note that the old wooden chute is still in place.*

Escola print
Humboldt State University

*Robertson raft under construction at Noyo in the early 1890s. At least three of these rafts were built here, but they were not successful. They were built in cradles on ship ways, and much difficulty was encountered in launching them. Later, on the Columbia River, a method was devised for assembling them in a floating cradle, and nearly one hundred fifty of these so-called "cigar" rafts were towed to California ports.*

E. Compton photo
Nannie M. Escola

*The landing at Rockport, with the schooner "Venture" loading at the wire chute. The top of this offshore rock has been blasted away to provide a storage area for the landing, and a suspension bridge connects it with the shore.*

Robert J. Lee Collection
Nannie M. Escola

**Loading under the wire chute at Noyo.**

Union Lumber Company photo
Nannie M. Escola

*Robertson raft leaving the Noyo River, heading for San Francisco Bay. One such raft broke up on arrival, where it raised havoc with shipping in the bay.*

ally used to serve harbors not accessible to the larger steam schooners. Because of the fumes that permeated them from the primitive gasoline engines, these ships came to be known as "skunks." Later, loggers throughout the West applied the term to their gasoline-powered rail cars, and for the same reason. Today the term is perpetuated in the popular passenger trains operated between Fort Bragg and Willits by the California Western railroad.

Nannie M. Escola

*Westport Landing in 1920, with a steam schooner loading from a wire chute. The age of steam has brought more sophisticated machinery, but the rocks remain as forbidding as ever.*

# CHAPTER SIX

# THE DONKEY ENGINE

AS THE POPULARITY of the steam engine spread, it was inevitable that it should be discovered by the loggers. The first man to recognize its potential was John Dolbeer, a partner in the Dolbeer & Carson Lumber Company of Eureka. A patent for his "logging engine" was issued on April 18, 1882.

The steam winch was nothing new. It had been employed by industry for some time: handling hoists in the mines, moving earth, operating pile drivers, and doing a multitude of other jobs. But until John Dolbeer conceived his "donkey," it had not been tried in the woods. There were several good reasons for this, and perhaps the most logical was the fact that until logs of the weight and dimensions of the redwoods had been encountered, there was no great need for mechanical power. In New England and the Lake States, where logs were fairly small and landings seldom lasted more than a week or two, horses and oxen did the job easily. The logs were decked beside a stream or railroad, and within a few days the crews moved on. A donkey mounted on its heavy sled, and with its need for fuel and water, would have been a nuisance.

But here in the West, where the timber was large and grew thick on the ground, a setting could last for months. Here the donkey proved its worth, and within a very short time Dolbeer's machine was

*Bancroft Library*

*John Dolbeer was born in Epsom, New Hampshire, on March 23, 1827. He joined the gold seekers to Gold Bluffs, north of Trinidad, during 1851. In 1853 he joined with other men to organize the Bay Mill. William Carson bought a half interest in the mill in 1863, and in 1864 the Dolbeer & Carson Lumber Company was organized. John Dolbeer was an inventive genius who contributed to the advancement of logging techniques in the West. In 1863 he patented a mechanical-tallying device to total the footage cut by a mill in any given length of time. In 1881, he received a patent for his first donkey, followed in 1883 by an improved version. In 1883 he also received a patent for his logging locomotive. Dolbeer was the senior partner in the firm and took the responsibility for handling the sales. After 1866, he made his home in San Francisco, where he died in August of 1902, at the age of 75.*

aiding the bulls in operations from San Francisco Bay to British Columbia.

Oddly enough, when Dolbeer conceived his machine it bore little resemblance to the common hoisting engines. Instead, he turned to the shipboard steam winch, which was perhaps more familiar to him. From this source, too, came the name "donkey." The first design consisted simply of a shaft with a gypsy drum on each end, driven through gears by a steam cylinder. Steam was supplied by a small vertical boiler, and the entire machine was mounted on a sled. Its success was immediate, although it had some

*A. W. Ericson photo*

*An early Dolbeer donkey at work in the woods near Mad River for the John Vance Mill & Lumber Co.*

*A. W. Ericson photo*

*Rolling a large butt log into the skidroad with a Dolbeer donkey. Note the use of manila rope, before the advent of wire rope for logging.*

Dolbeer used for loading. Here wire rope is being used, which has been carefully coiled by the spooltender to avoid being tangled.

*A. W. Ericson photo*

Hammond Lumber Company

*Yarding with a Dolbeer.*

obvious shortcomings. Manila rope was used at first, and while the engineer operated the machine, a "spooltender" wrapped the rope around the gypsy for the pull. As the line came in, it was carefully coiled by the spooltender. A line horse then carried the end of the rope back into the woods for the next pull. A major problem with the early machines arose from the fact that the pull was from the side, which tended to swing the donkey out of line. Dolbeer soon corrected this weakness by switching from the winch to a vertical capstan, which allowed for a straight pull from any direction. So successful was this new design, it was still doing regular duty in the red-

woods for a number of operators until the 1920s.

The Dolbeer, itself, didn't replace the bull teams, but it freed them for work on the skidroads where their efficiency was greatest. The tedious and time-consuming chores of working the logs into the skidroads and loading them aboard the cars were taken over by the steam engine.

On December 25, 1883, John Dolbeer was granted another patent that carried his idea a step further. In this patent, his winch was incorporated in a small locomotive, which gave it additional mobility. The winch was mounted on the front of a geared locomotive and was

A. W. Ericson photos

*Left top: Setting up the donkey to start yarding on a new side. Middle: Rigging up to yard some oversized logs at the Vance operations on Mad River. Bottom: later version of the Dolbeer donkey, with the vertical capstan. It was this type of donkey that was most often found in the western woods. In this scene, the yarding is being done with manila rope.*

driven by steam from the boiler of the locomotive, itself. The locomotive was now capable of loading cars, as well as other jobs along the railroad. The logging locomotive also proved popular with the redwood loggers, and a number of them were soon in service. But loggers elsewhere failed to be impressed, and the little gypsy locomotives remained a peculiarity of the redwood region.

The success of the Dolbeer donkey inspired further efforts to replace the plodding bull teams on the skidroads. The first such experiment was tried by Captain Robert Dollar at the Wonderly Mill at Usal, north of Rockport on the Mendocino coast. He set up an elaborate set of hoist-

ing drums, powered by a large horizontal boiler set over a brick oven. It was not designed to be practical from an economic standpoint. Rather, it was an experiment to determine if the idea had merit. Loggers from throughout the region came to observe it in action. The results proved so impressive that the idea spread rapidly. These new machines were known as "bull donkeys," for they replaced the bulls on the skidroads. The huge drums carried a mile or more of line and could move trains of logs no bull team could ever handle.

Having reached this point of development, the redwood loggers became suddenly resistant to change. Loggers on the

*Swanlund photo*

*It took plenty of blocks and line to provide a little Dolbeer with the leverage required to work a big log into place.*

Captain Robert Dollar was born in Falkirk, Scotland, March 20, 1844. He migrated to Quebec with his family when he was fourteen and got a job as chore boy in the kitchen of a lumber camp. He worked his way up to lumberjack and soon went into business for himself. His first venture failed, and he and his partner wound up with debts totalling five thousand dollars. He worked for three years to pay it off. His second venture proved successful, and in 1882 he came to the United States. He began amassing large timber holdings in Western Canada, Oregon, and California. He soon added shipping to his lumber interests and developed a trade with the Orient. At the age of 80, he started his round-the-world shipping service with the famous "Dollar Line" ships. He celebrated his golden wedding anniversary in 1925. At the age of 88, he came down with a cold that turned into pneumonia, and on May 16, 1932 he died at his home in San Rafael, California.

San Francisco, July 31, 1924

The Timberman:

I recently came across a very old photograph showing the first donkey ever used in the redwoods on a long haul. We took an ordinary boiler and bricked it in in order to make the experiment. Can you imagine bricking in a boiler to use in the woods? It looks like a joke, but this was a start, and many lumbermen visited our operation to see how the experiment worked. I am sending the photo along to show how primitive it all was. You will note the waterworks we had – all barrels. I thought you would be interested to see this, as it was real pioneering. The engine was built by Murray Bros., of San Francisco, and it was used at Usal, in Mendocino County, in operations I was managing.

Robert Dollar

(Capt. Dollar had taken over the management of the Wonderly Mill at Usal, near Rockport.)

In this landing scene at the Vance operations, we see the bull donkey used for roading the logs and the Dolbeer used for loading.

Big turn of logs being brought in for the Excelsior Redwood Company at Freshwater.

EXCELSIOR REDWOOD CO
EUREKA, CAL.
10 FT. DIA.

*Fine view of an early bull donkey. These machines could keep a woodcutter busy supplying fuel.*

*Patent drawing of the Dolbeer Logging Engine.*

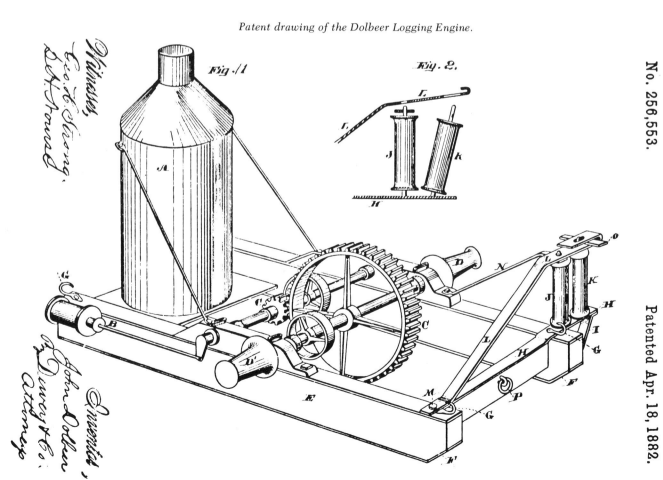

(No Model.)

No. 256,553.

J. DOLBEER.
LOGGING ENGINE.

Patented Apr. 18, 1882.

2 Sheets—Sheet 1.

Hammond Lumber Company

*Good view of a skidroad with a turn of logs, followed by the pig.*

Hammond Lumber Company

*Big road engine "California," built by the California Iron Works, at Eureka.*

Columbia River and Puget Sound far to the north, having accepted the donkeys with open arms, now began to refine and improve them. Within a few years they brought about a complete revolution in the industry. Huge machines were whisking the biggest logs out of the northwest woods like matchsticks, but the redwood logger clung tenaciously to his Dolbeer donkey and his tiny locomotives, fearful that this new power would only shatter his delicate timber and add to his wastage. It was not until the period of World War I that he began to accept the refinements on the machines that he had pioneered.

*Early logging scene in Humboldt County.*

A. W. Ericson photo

*John Vance was one of the early users of the large donkey for roading logs, more commonly called a "bull donkey." This one is unusual because of the wooden lagging used on the boiler. The little locomotive "Gypsy" is in the background.*

From B. H. Ward

*In the late 1860s Alexander Duncan was using this San Franciso built locomotive to handle his log trains on Austin Creek, in Sonoma County, near Duncan's Mills. The locomotive was affectionately known as "Mrs. Duncan's Teakettle."*

# CHAPTER SEVEN

# THE MENDOCINO RAILROADS

THE YEAR 1875 was significant in the history of logging on the northern California coast, for it was in that year that the steam locomotive made its appearance. In several locations, from Albion in the south to Arcata in the north, the woods began to echo to the sounds of the steam exhaust and the cry of the whistle. The tram roads, for the most part, had reached their maximum growth, and the efficiency of the steam railroad was drawing the attention of all elements of the population. Every community foresaw the day when the railroad would join it to the great centers of commerce, and every railroad endeavor brought the promise closer. Many of the logging railroads were projected over the mountains to a connection with some major rail system, though few carried the idea beyond the talking stage.

An important factor in the logger's advance to the railroad was the development of the locomotive industry in San Francisco. Almost as soon as the railroad

*Nannie M. Escola*

*Glynn & Peterson Lumber Company train at Delmar Landing, just south of Gualala, in Sonoma County in 1905. The three-foot gauge 2-4-2T once ran on the Park & Cliff House Railway in San Francisco.*

Nannie M. Escola
From M. Koch

*Gualala Mill Company No. 2 at the dump. The unusually wide gauge of 68½ inches is apparent in the picture.*

B. H. Ward

*Gualala Mill Company No. 1 was built in San Francisco by Miners Foundry & Machine Works about 1878, one of only three locomotives known to have been built by this concern.*

*Gualala Mill Company No. 2 from another angle. The wide gauge didn't keep her from turning bottom side up.*

*Two San Francisco-built locomotives of the Gualala Mill Company. They are No. 2, the "S. H. Harmon," and No. 1, the "W. B. Heywood," named for company officials.*

*Gualala Mill Company No. 3, one of two Baldwin-built engines, working in the woods.*

arrived in California, steps were taken to make the industry as independent of Eastern manufacturers as possible. Anything ordered through established channels had to be shipped around South America, a long and expensive trip that made expansion plans of the railroads uncertain. With the development of reliable suppliers in the San Francisco area, the new frontier became fairly independent of the rest of the country. By the 1870s, these plants were supplying equipment for the mainline railroads, as well as the small industrial locomotives ideally suited to logging.

At Caspar, the Caspar Lumber Company became the first to put a steam locomotive in operation in Mendocino County, attaining this honor by setting a second hand engine on the wooden rails of the old tram road. To support the locomotive, the rails were topped with strap iron, which quickly proved inadequate. Some forty-pound iron rails were soon acquired to replace the original rails. They had been salvaged from a wrecked ship, and required a good deal of work to get the kinks out of them, but they did the job.

Caspar Lumber Company was one of the few companies to cling to the hope of a mainline connection. The line grew slowly, but as the rails pressed eastward toward Willits, the dream grew brighter. However, by the time the goal was within reach, the incentive had vanished. A dwindling timber supply and the advent of the motor truck ultimately brought an end to the railroad operation — and the dream.

The first railroad operation was around Caspar Creek, close to the mill, at which time it was known as the Caspar Creek RR. A major extension soon carried the line north to Jughandle Creek. Within a short time, activity was transferred to Hare Creek by means of the famous Jughandle Creek trestle. The title was now

*Nannie M. Escola*
*The Stevenson bridge on the Elk Creek RR. of the L. E. White Lumber Company, near Greenwood.*

changed to the Caspar & Hare Creek RR. The year was 1884. It was 1903 before another major extension was indicated. The Hare Creek basin had been logged out, and the company was now prepared to move into the area drained by the South Fork of the Noyo River. To cross the divide, it was necessary to construct a tunnel nearly a mile in length.

With completion of the tunnel in 1903, another locomotive was added to the roster. This was a Climax, the first standard gauge, three-truck engine built by that company. The railroad was once more reincorporated as the Caspar, South Fork & Eastern RR., the title it retained until the end of service.

Following the purchase of the Climax, which proved somewhat of a disappointment, the company purchased a small Mallet from Baldwin. And later, when an additional locomotive was needed, a second Mallet was purchased. Both of these engines had been designed by Baldwin

especially for the road, keeping in mind the limitations imposed by the necessity of shipping them in over the wharf at Caspar. And, since Caspar remained committed to link and pin couplers throughout its lifetime, these were possibly the only Mallets to be so equipped in the United States.

In December, 1945, all operations were shut down by a strike, during which time the company reassessed their operations. With the resumption of logging, the changeover was made to trucks and the railroad was abandoned.

Another locomotive went to work at Rockport in 1875, and while this operation was marginal, many attempts were made over the years to make it a major producer. The last, and most ambitious attempt, was made by the Finkbine-Guild Lumber Company in the late 1920s. But, after a large expenditure of capital, and before a profit could be

*The Salsig store and post office about 1905. At this time, E. B. Salsig was manager of the L. E. White Lumber Company and the Greenwood Railroad. Frank Orr, standing under the post office sign, was storekeeper, timekeeper, and postmaster. The distinguished-looking gentleman standing beside the headlight is Tom Smythe, the company surveyor.*

*The Greenwood landing from an early post card.*

*L. E. White Lumber Company No. 1 pauses at the water spout. The "one spot" was a handsome little narrow gauge 4-4-0.*

*Later view of the landing at Greenwood. It has now been extended to the outer rock with the use of a wooden truss bridge.*

Nannie M. Escola

*Albion Lumber Company No. 3 working the branch up Marsh Creek, just south of Comptche. This was the former Albion River RR. No. 1.*

John E. Lewis
From Walt Casler

*Mendocino Lumber Company dump scene, contrived for the photographer in order to show the scope of the activity.*

Nannie M. Escola
M. Koch

*In many places the railroad clung precariously to cliffs above the sea.*

E. Compton print
Nannie M. Escola

*Mendocino Lumber Company tug boat "Maru," built to handle logs between the dump in Big River and the mill boom.*

Nannie M. Escola

*At the L. E. White Lumber Company dump, it was a thirty-five foot drop to the water.*

Nannie M. Escola
M. Koch

*Tallying ties at the Rollerville Landing, Point Arena operation of L. E. White, 1914.*

*Switching lumber cars at the L. W. White mill near Greenwood.*

*Moving the camp of Navarro Lumber Company, which operated off the farthest extension of the Albion branch, north of Boonville.*

B. H. Ward

*Gualala Mill Company No. 4 poses in front of the mill with a train of logs.*

Nannie M. Escola

*Another view of the L. E. White mill. Note the four-wheeled flat cars used for handling the lumber.*

John E. Lewis
From Walt Casler

*The Mendocino Lumber Company dump in Big River.*

Nannie M. Escola

*Unloading locomotive No. 202 of the Northwestern Pacific at the wharf at Albion. All of the locomotives and equipment for the Mendocino railroads had to come up from San Francisco by sea, but not all of the unloading facilities were as convenient as this one.*

*View of the mill and harbor at Albion. This was the third mill of the Albion Lumber Company, about 1900.*

*Albion River RR. No. 1 with some big logs.*

*Big mill of the Mendocino Lumber Company at the mouth of Big River.*

*Loading with the Lawson skyline system, often called the "Flyer." This system was developed by Davenport "Port" Lawson and is sometimes called the Port Lawson skyline. Lawson was superintendent for Goodyear Redwood Company, successor to L. E. White. It was a refinement of the McCanse Flyer, first set up by Sam McCanse, his predecessor. It was especially useful in spanning the deep and narrow gulches found along the Mendocino coast, and spans as great as eight thousand feet were used successfully.*

realized, the Great Depression of the 1930s put an end to it.

A narrow gauge railroad was incorporated in 1875 as the Mendocino RR. Company. This road was built up Greenwood Creek from the landing at Cuffey's Cove, just north of Elk, and was projected as a common carrier to serve several mills. In 1884 the road became the property of the L. E. White Lumber Company. Mr. White had built a narrow gauge road of his own in 1877 further north near Albion. This road was called the Salmon Creek RR., and it extended some eight miles up Salmon Creek from the wharf at the mouth of the creek. In addition to these two lines, Mr. White also operated a short narrow gauge line at Point Arena, which gave him a fairly wide spread system of narrow gauge railroads. Of these, the road in the Elk-Greenwood district was the most extensive. At one time called the Elk Creek RR., it was known during most of its lifetime as the Greenwood RR. It was operated by the L. E. White Lumber Company until 1916, when it was sold to

*Caspar Lumber Company at the mouth of Caspar Creek. A log from the chute has just hit the pond with a mighty splash.*

*Caspar Lumber Company dump with a log on its way down the chute.*

the Goodyear Redwood Company. It continued in operation until the late 1930s, although operations during the depression years were irregular.

The railroad at Gualala is worthy of note. Begun as a tram road, it was built to a gauge of 68½ inches, in order to allow a team of horses to walk between the rails. And when the road was converted to steam in the late 1870s, the gauge was retained. This was the widest railroad gauge used in the West, although some of the pole roads exceeded it. The original engines were supplied by San Francisco builders. Two later engines came from the Baldwin plant. The big mill burned in

*The Mendocino "two spot" with the woods crew.*

*Caspar Lumber Company No. 1, the "Jumbo," on the Jughandle Creek trestle.*

*Crossing a trestle on the Albion River RR.*

September of 1906, while negotiations were in progress for its sale, and despite the fact that machinery for a new mill was on the ground and a large tract of fine timber was available, the mill was never rebuilt. For many years the line operated sporadically, getting out ties and split material, until the final bankruptcy in 1922. During this time, the railroad owned four locomotives, all specially built to their unusual requirements.

Another important line was built out of Albion. First construction took place in 1881, when a horse tram was built. This was followed in 1885 by a standard gauge railroad up the Albion River, incorporated as the Albion River RR. This line was sold in 1902 and incorporated by the new owners as the Albion & Southeastern

*Mendocino Lumber Company No. 1, a steam dummy from the streets of San Francisco.*

*Caspar Lumber Company No. 2, "Daisy," in the woods with a ballast train.*

*Fort Bragg RR. No. 1, the "Sequoia," on one of the A-frame bridges that were to become a hallmark of the road.*

RR. It was sold once more in 1905, this time to the Fort Bragg & Southeastern RR. Company, which had been incorporated by the Santa Fe as part of a system being built in competition with the Southern Pacific. However, in 1907 the two big companies agreed to a joint ownership under the Northwestern Pacific RR. Company, and the line was transferred to that company.

Almost from its inception, the Albion Branch, as it was called, was projected to a connection with the main system in the neighborhood of Healdsburg. For one reason or another, this never came about, and in 1921 the line was leased to the Albion Lumber Company, which was then a subsidiary of the Southern Pacific. By 1929, operations in the valley had come to a standstill, and the lumber company ceased payments on the lease. The Northwestern Pacific received permission to abandon the line in 1930, but it was 1937 before it was finally dismantled.

At Fort Bragg, the Fort Bragg Redwood Company incorporated the Fort Bragg

*Nannie M. Escola*

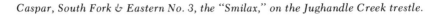

*Caspar, South Fork & Eastern No. 3, the "Smilax," on the Jughandle Creek trestle.*

*Caspar, South Fork & Eastern No. 4, the "Hercules," in front of the tunnel. Hercules was the first standard gauge three truck Climax built by that company and didn't prove as successful as had been hoped. She was later rebuilt into a two-truck engine.*

*Scene at the head of a Caspar Lumber Company incline. The building houses the lowering machine.*

*A log train on the Fort Bragg RR., headed by locomotive No. 3, drifts out of the mists on its way to the mill at Fort Bragg.*

*Caspar, South Fork & Eastern No. 7, the "Samson," on the Jughandle Creek trestle in 1925.*

*Early view of the Union Lumber Company plant at Fort Bragg.*

*Caspar, South Fork & Eastern No. 5. First of the Mallets, she was named the "Trojan."*

Nannie M. Escola

*Loading in the Caspar woods with the Lawson skyline system, commonly called a "Flyer."*

B. H. Ward

*California Western 2-6-2 No. 23
at Ft. Bragg.*

Horace Weller Collection
Nannie M. Escola

*Early operations in the Ten Mile River area, out of which came the Union Lumber Company.*

RR. Company in 1885 and started building eastward into the woods. The company grew and prospered, and in 1893 was reincorporated as the Union Lumber Company which continued to operate under this title until 1969, when the company was merged into the Boise-Cascade Corporation. In January 1973 the Georgia-Pacific Corporation purchased the entire properties of the Boise-Cascade, Union Lumber Region.

The railroad was reincorporated as the California-Western RR. & Navigation Company in 1905 and continued to push the rails eastward toward Willits and the Northwestern Pacific. At the close of 1911, the connection was finally achieved, and the California Western became the first and only Mendocino railroad to realize that elusive goal.

Passenger trains provided an important service from the beginning, as the Mendocino Coast remained a remote and inaccessible region. And even after the coming of the highway, the railroad still served areas having no other access to the outside world. But, as passenger traffic declined, the company turned to gasoline rail cars, quickly nicknamed the "Skunks," to handle the reduced business. As the years passed, the little yellow Skunks began to build an ever widening reputation. Today, the line is known officially by its more familiar name of

*Nannie M. Escola*

*The Glen Blair Redwood Company mill, as it looked in 1890. This plant, located just east of Fort Bragg, was later operated as a subsidiary of the Union Lumber Company.*

Nannie M. Escola
From M. Koch

This picture, taken in 1926, shows the equipment and shops of the Finkbine-Guild Lumber Company at Rockport. The headquarters of the company were in Mississippi, and the raw lumber was shipped there for finishing in their southern mills. The operation did not prove practical, and the plant was soon shut down.

Irvine & Muir used this primitive train at their operations at Irmulco, just west of Willits.

B. H. Ward

California Western train near Camp 10. No. 21 is a 2-6-2.

*Labbe photo*

*Motor car M-100, one of the famous California Western "Skunks," dozes at the station in Willits on a fall day in 1960.*

*Edward Freitas photo*

*Charles Russell Johnson, founder of the Union Lumber Company, or C. R., as he was known, arrived on the Mendocino Coast in 1882 at the age of 23 and purchased an interest in a small mill north of Fort Bragg. The capacity of the mill was very limited, and C. R. decided to build a much larger plant, choosing the area at Fort Bragg as the most practical location. With financial aid from his father, and other eastern lumbermen, he purchased the land around the present city of Fort Bragg and large timber tracts to support the operation. The new mill, using a band saw head rig, went into production in November 1885. From this beginning grew the giant Union Lumber Company, which, in 1905, absorbed three other lumber companies in the area. These were the Little Valley Lumber Company to the north, the Glen Blair Redwood Company to the east, and the Mendocino Lumber Company to the south. Upon his retirement in 1939, C. R. was succeeded by his son, Otis R. Johnson, and in February of the following year he died, at the age of 81. The company remained in the control of the Johnson family until it was merged into the Boise Cascade Corporation in January 1969.*

*Graves photo*

*California Western No. 17, one of several similar locomotives, commonly found in passenger service.*

*The Finkbine-Guild Lumber Company plant at Rockport.*

California Western RR., and its scenic route has become a mecca for tourists and railfans from around the world. Encouraged by the crowds of eager riders, the company has once again established passenger trains hauled by steam locomotives and dubbed the "Super Skunks." The faithful yellow Skunks are now held in reserve for the slack season.

The Georgia-Pacific plant remains as the lone lumber operation on the Mendocino Coast, healthy and active primarily because of the rail line that joins it to the nation. And the rail line, in turn, enjoys a healthy life of its own based on the product of the mill. That it also enjoys a reputation as one of the very few independent companies offering passenger service is due to the fortune of its location and foresight of Georgia-Pacific in promoting its attractions.

These are but a few of the railroads that served the Mendocino loggers. During the three quarters of a century that steam served to keep the lumber industry alive along the coast, there was scarcely a ravine that didn't shake to the rumble of passing log trains or echo the wail of the steam whistle.

*Humboldt State University*

*Builder's picture of the little Mattole Lumber Company No. 1. The remains of this locomotive rested in a stream bottom for many years until rescued by Henry Sorenson of McKinleyville, California. Henry has restored the engine and occasionally fires her up for a trip around the trackage in his yard.*

*John Cummings*

*Union Lumber Company Shay No. 2 was formerly the property of subsidiary Glen Blair Redwood Company.*

*Comprehensive view of an early railroad landing. The bull donkey is bringing a turn of logs to the rollway, while the Dolbeer, mounted to the left of the tracks, is waiting to load them on the cars. Additional skidroads are under construction in the foreground, one of which will carry the logs over the tail tracks at the landing.*

# CHAPTER EIGHT

# THE RAILROADS OF HUMBOLDT

BECAUSE HUMBOLDT Bay provided the best harbor on the northern coast, most of the activity in Humboldt County centered in that area. The bay, itself, was generally shallow, with two narrow channels surrounded by large tidal flats. Only in the immediate vicinity of Eureka did deep water extend close to the shoreline. But the main overland trails met the bay at its northern end, at the present site of Arcata, and here the mud flats extended a considerable distance to the nearest channel.

*Martha Roscoe*

*John Vance was born in Nova Scotia. Before heading for California with the gold seekers, he was a ship's carpenter. He arrived on Humboldt Bay in February of 1852 and became a leader in the development of that area. As a carpenter and millwright, he helped to convert the old steamer "Santa Clara" into the first successful sawmill on the bay for Ryan, Duff & Company. He later worked in the mercantile business, but soon turned to lumbering, buying a mill at the foot of G Street which had not been successful. He invested his profits in timberlands and eventually became the owner of thousands of acres of the finest redwood stands to be found in the county. By 1857 he had acquired a second mill and controlled the Eureka waterfront from F Street to J Street. Having logged off his timber back of Eureka, he moved north to the Mad River country and built the Big Bonanza mill on Lindsay Creek in 1875. At one time he nearly monopolized the Oriental lumber trade. He died on January 23, 1892, while serving his second term as Mayor of Eureka.*

*A. W. Ericson photo*

*The Humboldt Redwoods.*

A. W. Ericson photo

*In this view of the Big Bonanza mill, taken in the 1880s, the log train is headed by the "Onward," while on the lower track the little "Advance" is doing duty as the mill switcher.*

A. W. Ericson photo
Hammond Lumber Company

*The photographer on the left is preparing to capture on film company officials posing with a large log, while the "Onward" waits.*

A. W. Ericson photo

*An early landing in the Vance woods on Mad River. Two locomotives of the Vance's Mad River RR. are in evidence. On the left is shown the cab and tender of the "Advance," while No. 2, the "Gypsy," works the landing, which is located in the center of the logging camp.*

Henry Sorenson

*Landing scene of the 1890s. The bull donkey is mounted on a solid base, rather than a sled, and the unusual framework in front of it carries the leads for the mainline and haulback.*

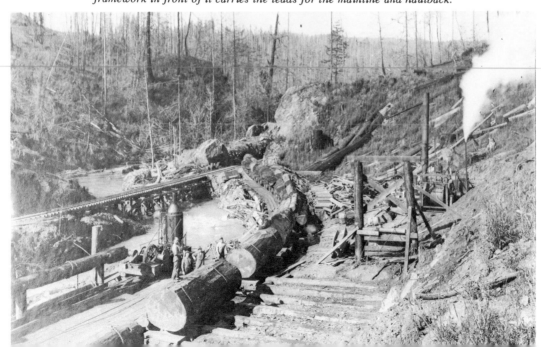

In order to create a port in the vicinity of Arcata, then called Union, it was necessary to build a wharf out to the ship channel. On December 15, 1854, the Union Wharf & Plank Walk Co. was incorporated for this purpose, and by late spring of 1855 had extended it two miles across the tide flats. To handle cargo on the long wharf, a railroad was constructed with wooden rails and utilizing a car drawn by a horse with the imaginative name of "Spanking Fury." This railroad was laid to the unusual gauge of 45¼ inches, the reason for which remains clouded in obscurity. One story has it that the only

available piece of shafting was cut into two equal pieces to provide the axles for the original car. But whatever the reason, the railroad was committed to this odd gauge for nearly ninety years. Having gone into operation early in 1855, the Union Wharf & Plank Walk Company laid claim to having been the very first common carrier railroad in California.

It was 1875 before steam came to the road, which was then reorganized as the Union Plank Walk & Railroad Company. The line was extended three quarters of a mile at that time to reach the Dolly Varden mill, built in 1872 by Isaac Minor and

*A. W. Ericson photo*

*Vance's steamer "Antelope" at the Mad River Slough landing in the 1880s.*

Hammond Lumber Company

*The mill at West Eureka (Samoa) was built by Vance's sons shortly after his death. This picture, taken in 1900, shows the property just before it was purchased by A. B. Hammond.*

A. W. Ericson photo
California State University, Humboldt

*The little "Gypsy" works a landing in the woods. The railing of heavy timbers, to the left of the train, was common in the area at this period. It provided a hold for the blocks used in loading the cars.*

A. W. Ericson photo

*Several large sections of redwood trees were shipped out of the Humboldt woods for exhibition at fairs and expositions. This one was known as "The Astor Cut," and the story is that John Jacob Astor bet a large sum of money that he could provide a table built from a solid slab of redwood to seat a banquet. This cut was shipped to London, where it was made into a table. It is now on display in a park in London.*

Humboldt State University

Using the gypsy to load the cars.

Henry Sorenson

*Hammond Lumber Company camp in the heyday of railroad logging.*

*Andrew Benoni Hammond was born in St. Leonard's, New Brunswick, in 1848. After coming to the United States, he spent a short time in Pennsylvania before heading for Montana in 1867, when he was 19. He went to Puget Sound by way of Portland, where he logged for a time on Hoods Canal for Pope & Talbot. After a year on the Sound, he returned to Montana, and in 1869 he went to work in the store of Bonner & Walsh. He later bought an interest, and in 1885 The Missoula Mercantile Company was organized to take it over. He retained an interest in this company throughout his lifetime. In the early 1880s, he formed the Big Blackfoot Milling Company and supplied most of the ties and timber for the westward expansion of the Northern Pacific. This operation was sold to Anaconda Copper Mining Company in 1898.*

*About this time, in the company of E. L. Bonner, he came to Oregon, where extensive purchases of timber were made. The two men also bought the Oregon Pacific RR. and made an agreement to build the Astoria & Columbia River RR. from Astoria to a connection with the Northern Pacific at Goble. Shortly thereafter, Mr. Bonner withdrew from the association, and Mr. Hammond turned to the Huntingtons of the Southern Pacific for his support. At this time, also, the Hammond Lumber Company was formed to take care of the extensive interests in sawmills and timber.*

*In addition to his other interests in the Astoria area, he formed the Columbia River Packers Association, and he retained ownership of the concern until 1925.*

*In 1900, the interest of the Southern Pacific drew him to the redwood forests of Humboldt County, where he began buying large tracts of timber. His first purchase included the Vance properties, which formed the base of his redwood empire, and the Eureka & Klamath River RR., to establish the Southern Pacific in northern Humboldt County. The Eureka & Klamath River RR. was reorganized as the Oregon & Eureka RR., and was leased back to the lumber company.*

*There were numerous additions to the property over the years, which occasioned several changes in corporate titles, until the entire operation became the property of the Georgia-Pacific Corporation in 1956 for a whopping $80,000,000.*

*A. B. Hammond died in January 1934, at the age of 85.*

Noah Falk. The locomotive, called the "Black Diamond," was built by the Eureka Iron Works with an upright boiler and four wheels driven through gears. Strap iron was laid on the wooden rails to accommodate the added weight. In 1876 the line was again extended, this time to reach the Jolly Giant mill at the northern edge of Arcata. This mill had been built in 1874 by Minor and Falk. The railroad provided service to Eureka by means of a small sidewheel steamer from the wharf.

In 1878 the name of the railroad was changed once more. It became the Arcata Transportation Company, and the rails were extended to a mill owned by Isaac Minor at Warren Creek.

Finally, in 1881, it became the Arcata & Mad River RR., and the strap iron rail was replaced with 35 pound T rail. It was now a full-fledged railroad. Affectionately referred to as the "Annie & Mary," it still provides service to the Simpson mill at Korbel. Most of the narrow gauge rails were changed to standard gauge in 1925, including all of the woods lines, but the last narrow gauge remnants were not scrapped until 1942.

Because the bay was surrounded by marshy flats, the loggers resorted early to the use of tram roads to reach the water with their logs. Within a few years after the opening of operations on the bay, many miles of these roads were in use.

In 1874, John Vance founded the John Vance Mill & Lumber Company and started construction of the Big Bonanza Mill on Mad River, at what is now called Essex. To provide access to the mill, he built five miles of standard gauge railroad from the mill site to tidewater on Mad River Slough, to the west of the Arcata wharf. In the magic year of 1875, he put a steam locomotive to work on his new line, then known as Vance's Mad River RR. Lumber was transferred to ships from the terminus on the slough by means of barges.

The road continued to extend its operations northward into the timber, handling

*A. W. Ericson photo*

*The Oregon & Eureka RR. No. 11 headed for Samoa with a long train from the woods around Little River. The locomotive was constructed by the Hammond company in their own shops at Samoa in 1910.*

Among the Humboldt Redwoods, California
A. W. Ericson, Foto. Registered

The Jolly Giant mill of Isaac Minor and Noah Falk at the north edge of Arcata as it appeared about 1876. Barely visible at the edge of the timber is the Union Plank Walk & RR. Co. No. 1 on wooden rails.

B. H. Ward

This little tank engine was No. 1 on the Arcata & Mad River RR. She was turned out by H. K. Porter & Company in 1881. Named the "Arcata," she was originally numbered 2, but when the old "Black Diamond" was retired she was renumbered.

*In this view the "Onward" crosses overhead, while the "Gypsy" works a woods spur below. The title of this road saw many changes and was most likely to have been called the Eureka & Klamath River RR. at this time. With the acquisition of the road by A. B. Hammond, the "Onward" was sent north to work on his railroad at Astoria, Oregon.*

*The Arcata & Mad River RR. No. 3, the "North Fork," trundles her train across the Mad River bridge.*

*The Arcata & Mad River steamer "Alta" provided a connection with the city of Eureka.*

logs to the mill, as well as the lumber. On December 26, 1891, John Vance sold his interests to his sons, who on January 6, 1892, transferred the railroad to their Humboldt Bay & Trinidad Logging & Lumbering Company. The railroad was renamed the Humboldt Bay & Mad River RR.

In 1893 the sons started construction of a new mill at Samoa, and plans were made to extend the railroad down the peninsula to the new mill. On January 12, 1896, the road was once again reorganized as the Eureka & Klamath River RR. It was now planned to serve Eureka by means of a ferry from the new terminus at Samoa. And, as indicated in the new title, northern extensions were projected to the Klamath River. Work was also begun on an extension into the town of Arcata, and ultimately around the bay to Eureka.

Also, in the year of 1875, a company known as the South Bay RR. & Land

*A. W. Ericson photo*

*Superintendent V. Zaruba, with his inspection car on the Arcata wharf, June 1, 1893.*

*A. W. Ericson photo*

*The tug "Mary Ann" helps a well-laden schooner toward the open sea from the Arcata wharf in 1891.*

*Isaac Minor was born in Pennsylvania on April 8, 1830. As with so many others, he came West with the early gold rush. He arrived in Humboldt County in December of 1853. He was one of the first to engage in logging in the Arcata area. In 1872, with Noah Falk, he built the Dolly Varden mill two miles north of Arcata. The following year the two men built the Jolly Giant mill in a gulch behind the present site of Humboldt State University at the north edge of Arcata. In 1875 he sold his interest in these mills to his sons and moved north to Mad River, where he continued his lumbering activities. He also had interests in a bank, a granite quarry, creameries, shipping, farms, stores, and timber land. He became one of the wealthiest men in California. In 1902 he sold twenty thousand acres of redwood timber in Del Norte County, which included the timber lying in Oregon. In 1908, at the age of 80, he married a second time, taking as his bride a woman fifty years his junior. In 1910, he sold eleven thousand acres of timber in the vicinity of Yosemite National Park, being one of the few lumbermen to have interests in both redwood regions. He died at Arcata on December 11, 1915, at the age of 87.*

Company built five miles of railroad up Salmon Creek from the southern end of Humboldt Bay to serve their Milford Mill and Lumber Company. By 1880, this area was cut out and the railroad was transfer-

red to Freshwater Slough as the Humboldt Logging Ry., where it ultimately came into the possession of the Excelsior Redwood Company. Further north at Trinidad, the Hooper Brothers were operating the Trinidad Mill Company, with a tram road connecting the mill with the wharf. This was one of the early tram roads in Humboldt County. It was converted to a standard steam railroad by the Hoopers, possibly as early as 1875.

*A. W. Ericson photo*

*Arcata & Mad River passenger train, headed by locomotive No. 5, the "Blue Lake," crossing the North Fork of Mad River about 1901.*

A. W. Ericson photo

*The plant of the Minor Mill & Lumber Company at Glendale, near Mad River, as it appeared in the late 1880s. This operation was served by the narrow gauge rails of the Arcata & Mad River RR.*

A. W. Ericson photo

*The Arcata & Mad River roundhouse at Arcata about 1890. The engines, left to right, are No. 3 "North Fork," No. 4 "Eureka," No. 5 "Blue Lake," and No. 2 "Arcata."*

G. W. Miller photo
Amos R. Tinkey

*The "Blue Lake" was a graceful little 2-4-0, shown here at Blue Lake.*

Humboldt State University

*Minor Mill & Lumber Company operated a plant at Glendale, west of Blue Lake, and a logging railroad served by this little gypsy engine ran north into the timber.*

The Arcata & Mad River engine house at Korbel in the early 1900s. The engines have now been renumbered, and No. 4, the "Blue Lake," is at the left. The new number 5, built by Baldwin in 1901, is named the "Hoopa." On the right is the "North Fork," now numbered 2.

Humboldt State University

Little locomotive used in the woods by Humboldt Lumber Mill Company, Korbel, owners of the Arcata & Mad River RR. Being a gypsy engine, like many others in the area, she was called the "Gypsy." She was built in San Francisco by Marshutz & Cantrell, as attested by the builder's plate at the base of the cab.

*The logging railroad of Minor Mill & Lumber Company connected at the mill with the Arcata & Mad River RR., and was, therefore, built to the same 45¼ inch gauge. The spool in the foreground carries the line used in loading with the gypsy drums on the locomotive.*

*A. W. Ericson photo*

*The Northern Redwood Lumber Company mill at Korbel, in the early 1900s.*

*In the final days of steam on the Annie & Mary, No. 11 is bringing a train down from the mill at Korbel. The 11 was an Alco 2-6-2, built in 1925, which was acquired by the A&MR in 1953.*

*Early locomotive built to the Dolbeer design. Note the wooden frame and solid drivers. This may have been pictured on the operations of Dolbeer & Carson Lumber Company on Humboldt Bay.*

**The steamship "Iran" loading at the Arcata wharf in the late 1890s.**

*A. W. Ericson photo*

*Mill of the Riverside Lumber Company at Korbel about 1890. This was one of the mills included in the merger which formed the Northern Redwood Lumber Company.*

*B. H. Ward*

*Two Heislers of Northern Redwood Lumber Company in the woods at Korbel.*

*View of Trinidad in the late 1870s, showing the Hooper Brothers mill and the railroad leading to the wharf.*

Dr. S. R. Wood photo
G. B. Abdill

*The Trinidad Mill Company "Sequoia" at work near Trinidad about 1880. This little Baldwin-built tank engine went to the Bucksport & Elk River RR. as their No. 2, "Trinidad," where she was finally scrapped in 1934.*

A Dolbeer locomotive at work near Humboldt Bay.

Arcata & Mad River No. 2, as she appeared at Riverside in 1925.

Loading shingles at Houda's Landing, near Trinidad, in the late 1880s.

*David Evans was one of the outstanding men of the redwood industry. He was born in Carmarthensire, Wales, in 1838, and came to Humboldt County in 1857, at the age of 19. He was well-known for his inventions, his ability as a millwright and his skill as a timber cruiser. Among his most useful patents was one for a third saw, which enabled a mill to cut oversized logs without having to split them first. He received this patent in 1869. Another, received in 1892, covered the system of snatch blocks used on the turns in skidroads to keep the proper lead on the line. During his lifetime, Mr. Evans was associated with the Occidental Mill, the Milford Mill, the Dolbeer & Carson Lumber Company, the California Redwood Company, and the Excelsior Redwood Company. He was very popular and served two terms as Mayor of Eureka, defeating William Carson for the office. He died July 12, 1901, at the age of 62.*

In 1883, the ill-fated California Redwood Company was formed to buy up all the mills and timber holdings in the area around Humboldt Bay. Included in this organization were two large mills on the bay and the Hooper interests at Trinidad. Some one hundred thousand acres of timber backed up the milling operations. Both the Trinidad railroad and the Humboldt Logging Ry. became the property of the California Redwood Company, while the Bucksport & Elk River RR. came partially under their control.

Operations of the California Redwood Company were suspended in 1886, when government investigations revealed fraudulent dealings in the acquisition of much of the timber. As a result, the Hooper Brothers took over one of the mills on Gunther Island, where they set up the Excelsior Redwood Company. The Humboldt Logging Ry. became part of this operation, and the earlier operations at Trinidad were closed out.

A. W. Ericson photo

*The "Emily" at the Trinidad wharf, 1890.*

*Excelsior Redwood Company train from the Freshwater woods.*

*Sailing vessels loading at the Excelsior Redwood Company mill on Gunther Island in 1890.*

*The Humboldt Logging RR. No. 3 from the Freshwater camp of the Excelsior Redwood Company with a load of huge logs.*

*The Timberman*
*Oregon Historical Society*

*Noah H. Falk was born in Clarksville, Pennsylvania, on June 11, 1836. When he was 12, the family moved to Ohio, but in August 1854 he arrived in San Francisco, where he got a job helping to build the original mill at Albion. On August 10, 1862, he was married in Seattle. In 1869 he helped establish the Janes Creek mill, just north of Arcata, which operated until 1888. In 1872 he joined with Isaac Minor in building the Dolly Varden mill at Arcata, and the following year the two men built the Jolly Giant mill nearby. In 1882 he started construction of a mill on Elk River, where the town of Falk was established. This was the first band mill in Humboldt County, and Noah Falk insisted that it be equipped with steel wheels over the strong objections of the manufacturer, who had always used wooden wheels. It proved to be a complete success. To serve this mill, the company incorporated the Elk River RR., which was soon sold to Dolbeer & Carson and renamed the Bucksport & Elk River RR. In 1920, at the age of 84, he sold his interest in the mill and retired. And in 1922, he and his wife celebrated their 60th wedding anniversary. He died at his home in Arcata on March 10, 1928, at the ripe old age of 92.*

*A. W. Ericson photo*

*Three trains pose for the photographer in the Excelsior camp at Freshwater.*

*Unusual articulated locomotive built for the Bayside Lumber Company in 1910 by the Eureka Foundry Company and numbered 1.
It is said she proved something less than satisfactory, as the cylinders, which were bolted to the frame, had a habit of working loose.*

*Flanigan, Brosnan & Company No. 2 at the dump on Gannon Slough, near Bayside, on Humboldt Bay. This company logged along
the north side of Jacoby Creek and eventually became the Bayside Lumber Company. This photo was taken just before the turn of
the century.*

*Log train of the Bucksport & Elk River RR. at the dump at Bucksport, headed by No. 1, an 0-4-2 built by Baldwin in 1884. The mill in the background was owned by Holmes-Eureka Lumber Company.*

*Flanigan, Brosnan & Company trains headed by locomotives No. 2 and No. 3. The 3-spot has a train of shingles from the mill at Bayside. In 1900, the Bayside Mill & Lumber Company was formed to take over this operation. And in 1905, it was sold once more to a group of Eastern buyers, who incorporated the Bayside Lumber Company.*

Jack Slattery

*McKay & Company operated the Occidental mill for many years and logged along Ryan Creek at the eastern edge of Eureka.*

Humboldt State University

*Early rod engine built by Globe of San Francisco. This locomotive was said to have arrived in Humboldt County in 1875. The winch on the front deck is powered by a separate set of cylinders, and it is possible that this was the inspiration for Dolbeer's patent.*

A. W. Gilfillan photo

**The Elk River mill at Falk in 1888.**

*Lester Holmes*

*The sawmill of Elk River Mill & Lumber Company under construction at Falk in 1883. This was the first mill in Humboldt County to be equipped with a band saw.*

*Gus Haggmark*

*No. 2 of the Elk River Mill & Lumber Company at Falk. This little 0-4-4T was a Baldwin product of 1903.*

*Henry Sorenson*

*Elk River Mill & Lumber Company No. 1, the "Falk," as she appeared early in her career at Falk. The plate of her builders, Marshutz & Cantrell, appears at the base of the cab. In 1936 the engine was donated to Fort Humboldt, in Eureka, where she is still on display.*

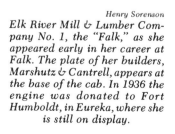

Miller-Freeman Publications

*Dolbeer & Carson No. 2 passing the Elk River shops with a train.*

Jack Slattery

*Baldwin locomotive "Sequoia," which was exhibited at the Lewis & Clark Exposition in Portland in 1905. It was purchased by Dolbeer & Carson and put to work on their Humboldt Northern RR. as their No. 2.*

B. H. Ward

*McKay & Company's 1 Spot was turned out by Globe of San Francisco.*

*Loggers were an inventive lot and were constantly experimenting with new ideas, not all of them entirely practical.*

*Builder's photo of McKay & Company's No. 2, turned out by H. K. Porter Company in 1904. It was named for James Loggie, brother-in-law of Allan McKay. McKay died in 1888 and Loggie took over the management. The locomotive was later rebuilt to a 2-4-4T.*

*William Carson was born in New Brunswick in 1825. He came to California in search of gold in 1849 and worked in the mines during the summer of 1850. Because food was scarce, he and his companions decided to spend the winter on Humboldt Bay, where game was plentiful. Martin White was building a small sawmill on the bay, and the group contracted to supply the logs. In November of 1850 they located their camp on Ryan*

In the early 1880s, various interests began to plan a large operation on the Eel River at a place called Forestville, now known as Scotia. This became the Pacific Lumber Company plant, which developed into one of the major concerns in the redwood industry. The location of the sawmill was some distance from a port, and a railroad was necessary before operations could begin. In 1882 the company incorporated the Humboldt Bay & Eel River RR. and grading was started from the bay. However, John Vance, William Carson, and others, had incorporated the

*Slough, where William Carson and Jerry Whitmore were supposed to have felled the first redwood tree for sawlogs in Humboldt County. In the summer of 1854 he leased the Muley Mill in Eureka, where he worked as sawyer. And that fall he shipped the first full cargo of redwood lumber from Humboldt Bay. In 1863 he purchased a half interest in the Bay Mill in Eureka, and in 1864 the Dolbeer & Carson Lumber Company was formed. Until its demise in 1950, this company was an important factor in the redwood industry of Humboldt County. William Carson built his home near the site of the mill and supervised the operations. In 1885 and 1886 he constructed the famous Carson Mansion, which remains today one of the outstanding monuments of Victorian architecture in the United States. He died in Eureka on February 19, 1912, at the age of 87.*

*Miller-Freeman Publications*

*Dolbeer & Carson No. 3 with a load of redwood logs from the Elk River country.*

Eel River & Eureka RR. to build south to the Van Duzen River in the same year, and the Pacific Lumber Company arranged for this company to handle their lumber trains. In 1885, they built their own line from Alton Junction, on the Van Duzen River, to the mill site. Logging extensions continued the road some distance south along the Eel River from Scotia.

In 1900, A. B. Hammond appeared on the scene. He bought out the Vance interests, including the Eureka & Klamath River RR. He was already deeply involved in railroad building in Oregon, where he was closely associated with the Huntington interests of the Southern Pacific. It was at this period that the Southern Pacific and the Santa Fe were gathering their forces to do battle for the lumber trade from the Humboldt area. In 1903, Southern Pacific acquired the California & Northwestern, which had built north from San Francisco Bay to the Willits area. And in an attempt to block the Santa Fe in the north, the Southern Pacific acquired the Eureka & Klamath River from Hammond. At this time the

A. W. Ericson

*The world-famous Carson mansion in Eureka. This magnificent home was built by William Carson in 1885-86, and remains today one of the outstanding examples of Victorian architecture in the United States. This photo was taken in 1892.*

*Peter J. Rutledge, born in New Brunswick, Canada, in 1874, came to Eureka with the family on October 7, 1883, and at the age of 15 went to work in the woods. He joined the Dolbeer & Carson Lumber Company in 1898, where he remained for 38 years, advancing from mill foreman to General Superintendent. He died in Eureka in 1960 at the age of 85.*

name was changed to the Oregon & Eureka RR., and the road was leased back to Hammond.

Meanwhile, in May, 1903, the Santa Fe formed the San Francisco & Northwestern to handle their interests in the redwood battle. They immediately purchased the Eel River & Eureka RR., as well as several smaller companies struggling to meet the competition of the Hammond lines around Humboldt Bay. Santa Fe also acquired the Pacific Lumber Company lines up the Eel River. Hammond made a bold attempt to acquire control of the Pacific Lumber Company, but was unable to get more than forty-eight per cent of the stock. With the sale of the railroad to the Santa Fe, Hammond disposed of his interest in the Pacific Lumber Company.

*San Francisco & Northwestern No. 5, a Baldwin product of 1886, at work at Happy Camp near Holmes.*

To reach San Francisco Bay, the Santa Fe had incorporated the Fort Bragg & Southeastern, which acquired the Albion & Southeastern road in Mendocino County. However, they found themselves effectively blocked by the California & Northwestern, and the battle soon became a stalemate.

Finally, in January, 1907, the Santa Fe and the Southern Pacific joined forces to form the Northwestern Pacific RR. In the fall of 1914, a through line was completed from Humboldt Bay to the San Francisco Bay port of Tiburon. Now, for the first time, the redwood mills were able to ship by rail as well as by sea. Ultimately, the Santa Fe sold its interest in the Northwestern Pacific to the Southern Pacific and retired from the redwood scene.

*The Pacific Lumber Company*

*Simon Jones Murphy, originally from Detroit, acquired control of the Pacific Lumber Company in 1905, and since that time the company has remained in the control of the family. Although Simon Murphy died two months before completing the transaction for the company, his sons and associates have made it one of the outstanding companies in the region.*

*The Pacific Lumber Company*

*Pacific Lumber Company No. 1 on a passenger train at Scotia.*

*Early picture of Pacific Lumber Company No. 2 at work in the woods near Scotia.*

*San Francisco & Northwestern RR. No. 0, the passenger train on the Carlotta branch, affectionately called "The Dummy."*

*View of Scotia and the big mill of The Pacific Lumber Company about 1902.*

*Pacific Lumber Company train
on Stitz Creek trestle in 1895.*

*Pacific Lumber Company No.
21, a classic Rogers 4-4-0.*

*Pacific Lumber Company No. 1
is dwarfed by a big log.*

*The Pacific Lumber Company*

At the Eureka & Freshwater RR. shops after being taken over by The Pacific Lumber Company. The little Excelsior Redwood Company "Deuce" now sports a tender. This is the only known picture showing a locomotive lettered for this line. The two larger engines, the No. 24 and No. 27, are Pacific Lumber Company engines.

*The Pacific Lumber Company*
*Pacific Lumber Company train in the woods.*

*The Pacific Lumber Company*

Locomotive No. 25 trundles a train across the Holmes-Shively bridge over Eel River in 1910.

*Albert Stanwood Murphy, grandson of Simon Murphy, assumed the presidency of the Pacific Lumber Company in 1931 and guided the company through the years of the Depression. He was made Chairman of the Board in 1961, and died in April, 1963.*

*Pile driver at work on a trestle for The Pacific Lumber Company.*

*Camp of The Pacific Lumber Company in the Freshwater operations. This is a railroad camp, and the camp cars are arranged in orderly rows on the sidings.*

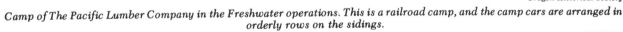

# CHAPTER NINE

# DEL NORTE

THE FIRST LUMBERING operation in Del Norte County appeared in 1853, when F. E. Weston set up a small mill in Crescent City to cut for the local market. The logs were hauled in on two large wheels twelve feet in diameter.

The first large mill was built by local businessmen in 1870. It was located on Lake Earl, north of Crescent City. It was soon sold to outside interests, who continued to operate it. It burned in 1890, but operations were soon resumed.

Not long after the mill was built on Lake Earl, Hobbs, Wall & Company built another mill at Crescent City. In the late 1880s, they built a railroad to their timber north of town. The road was incorporated as the Crescent City & Smith River RR., and by September of 1889, Smith River had been bridged. Soon passenger trains were running to the town of Smith River.

In 1903, Hobbs, Wall & Company purchased the sawmill, timber, and railroad of the Crescent City Wharf & Dock Company at Lake Earl, leaving the company as the only large operator in the county.

*Dick Childs*

*A Hobbs, Wall & Company Shay poses high atop a bridge of cribbed logs.*

*Dick Childs*

*Little Forney locomotive handles a log train for Hobbs, Wall & Company in the flat country north of Crescent City.*

*Dick Childs*

*Hobbs, Wall & Company train near Crescent City.*

*D. S. Richter*

*Camp buildings of the Brookings Timber & Lumber Company beside the Chetco River in southern Oregon.*

There were a number of railroad developments projected, which envisioned connecting the port of Crescent City with a major rail line in the interior, and their projectors all hoped to use the rails of the Crescent City & Smith River RR. as their access to the harbor. However, little was ever done to implement the plans, and eventually Hobbs, Wall removed the rails in the northern party of the county.

Around 1915, with the operations completed to the north of Crescent City, Hobbs, Wall turned their attention towards tracts lying to the south of town. The country in this area was more rugged, and the company switched to geared engines for the woods operations. However, still hoping for a connection with the rest of the United States, and noting the progress from the south being made by the rails of the Northwestern Pacific, they incorporated this new extension as the Del Norte Southern.

But despite all the plans for connecting rails, Del Norte County remained isolated. Finally, in the early 1940s, the company finished its operations and all of the rails were removed.

In 1912, another large concern made its appearance. The Brookings Lumber & Box Company, which had been operating in the San Bernardino Mountains east of Los Angeles, was closing out its activities in that area. Looking for new worlds to conquer, the Brookings family turned to the southern Oregon coast. They bought

*Dick Childs*

*Little Hobbs, Wall engine switching the yards at Crescent City. She was turned out by Globe of San Francisco.*

large tracts of timber along the Chetco River and laid plans to develop a landing near Harbor. While the timber was primarily Douglas fir, it also included what redwood stumpage lay within the borders of Oregon.

In June of 1913, the Brookings Timber & Lumber Company was incorporated, with a capital stock of $1,500,000. Work was begun on a large sawmill, and the town of Brookings was laid out across the Chetco River from Harbor. By 1914 the mill was in operation, and a logging railroad was bringing fir logs down from the Chetco area.

The Brookings operation was distinctive in that it was the only landing on the coast shipping fir lumber, and it soon found itself at a disadvantage in the very competitive fir market. By the summer of 1916 the big mill shut down.

A group of Wisconsin timbermen had been investing heavily in Del Norte timber, beginning with an original purchase of twenty thousand acres from Isaac Minor in 1902. Operating under the title of The Del Norte Company, they had been expanding their holdings and biding their time. Now, seeing an opportunity to realize on the investment, they came to the rescue of the Brookings operation.

In January 1916, the California & Oregon Lumber Company was formed to succeed the Brookings Timber & Lumber Company, and the capital was raised to

*Miller-Freeman Publications*

*The California & Oregon Lumber Company poses the No. 5 on the Chetco River bridge, an imposing structure they were justly proud of. The mill and wharf were located beyond the jutting point of land in the distance.*

five million dollars. The sawmill and railroad were rehabilitated, and in the spring of 1917 work was begun on the wharf, and the railroad to serve it.

The company continued to cut fir from the Chetco camps, while plans were developed for the switch to redwood. In the spring of 1922, the first redwood trees were cut in a tract on Rowdy Creek, just east of the town of Smith River. A railroad was under construction south from Brookings. It presented no difficulties, except for the impressive span across the Chetco River at Harbor. This bridge, constructed largely of wooden piles, was 2140 feet in length.

By September 1922, logs from Smith River were being taken to the mill at Brookings. Some consideration was given to building to a connection with the Hobbs, Wall & Company line, now five miles distant from Smith River, and building a redwood mill on Lake Earl, which would leave the Brookings mill free to cut fir. However, as time passed, it became more obvious that the big mill was unable to compete in the fir market. Operations were concentrated on timber from the Smith River district, and no more fir was cut. But the redwood operations were scarcely more profitable than the fir, and in the summer of 1925 the big mill cut its last log.

In the year that followed, much of the machinery was removed, and the railroad was torn up. Several vain attempts were made to operate on a smaller scale, but nothing was ever salvaged from the re-

*Dick Childs*

*California & Oregon Lumber Company train headed for Brookings, Oregon, with logs from Smith River, in Del Norte County, California.*

The Timberman Collection
Oregon Historical Society

*The big mill at Brookings, as seen from the ocean side.*

D. S. Richter

*Wharf scene at Brookings, Oregon.*

mains. Ironically, Brookings today is a
prosperous and growing community with
an economy based largely on a healthy
lumber industry.

*Dick Childs*

*Wharf of the California & Oregon Lumber Company at Brookings, Oregon. This marked the most northerly extension of the redwood belt and was the only such landing north of the California border. This company built a substantial plant and town here in 1916, but the operation was short-lived. In this scene, a Shay is taking out flat cars piled high with lumber for loading on the steam schooner tied up at the impressive wharf.*

*The modern day chopper uses a gasoline-powered chain saw, but little else has changed. The driving boards and staging are the same, and the trees still come big.*

*Arcata Redwood Company*

*Howard A. Libbey was born and raised in Eureka. In 1916 he went to work for the Little River Redwood Company and became general manager. The company was merged with Hammond Lumber Company in 1931, and Mr. Libbey retired. In 1934 he joined Hobbs, Wall & Company and the following year was appointed Vice President and General Manager at Crescent City. In 1936 he retired once more. With the formation of The Arcata Redwood Company on July 15, 1939, Mr. Libbey became the first president of that company. He remained in that position until his final retirement in 1967. He also served as president of the California Redwood Association for many years.*

*Swanlund photo*

*Sighting the fall with a gunstick. The undercut has been sawn with a chain saw and bears little resemblance to the massive undercuts once chopped by hand.*

*Swanlund photo*

*With the undercut finished, the chopper starts the backcut.*

*Timmmberrr! As the big tree starts its downward course, choppers Ted Williams and Wayne Miller scramble to safety. Scene is on the Big Lagoon operations of the George-Pacific Corporation, formerly Hammond Lumber Company.*

On the ground at last, the big tree dwarfs the choppers. The picture was taken at the Big Lagoon operations in 1958.

Swanlund photo

Setting chokers on a cat show.

*Yarding with a cat and arch at Big Lagoon, August 1956.*

Big Lagoon, August 1956.

Swanlund photo

*Just as the donkey came along to displace the bull teams, the cats displaced the donkeys in the 1930s. Today, however, with the growing concern for the ecology, the cat has lost favor. Loggers are turning to new methods that are less destructive to the terrain, and the skyline is making a comeback.*

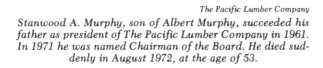

*Stanwood A. Murphy, son of Albert Murphy, succeeded his father as president of The Pacific Lumber Company in 1961. In 1971 he was named Chairman of the Board. He died suddenly in August 1972, at the age of 53.*

*Probably the heaviest loader ever built. This machine was turned out by the Washington Iron Works of Seattle to the order of The Pacific Lumber Company. Basically, it comprised the duplex loading unit from a tower skidder. However, it proved impractical, due to its great weight.*

Swanlund photo

*The old and the new at The Pacific Lumber Company.*

*Affiliated with Hammond Lumber Company and successor Georgia-Pacific Corporation for 45 years, A. F. "Bud" Peterson retired in December 1971. Born in Kerhoven, Minnesota, he left home in 1924, at the age of 17, and went to work in the woods of Alberta. A year later he paused long enough in Humboldt County to work half a day for The Pacific Lumber Company. He moved on to Sonora and then back to the Canadian mine fields before returning once more to Humboldt County in 1926. From that time on, he remained with Hammond, working up to production superintendent in 1970.*

*Swanlund photo*

*Crossing the Eel River on the private truck road of The Pacific Lumber Company.*

*Loading some big ones with a crotch line at Simpson Timber Company. The power is supplied by the cat-mounted drums in the background.*

*Hammond logs travel the coast highway near Big Lagoon.*

*Topping a spar tree in the 1930s for the Holmes-Eureka Lumber Company.*

*Hammond truck landing at Big Lagoon.*

*Tightening up the binders before heading out on the highway.*

Swanlund photo

*Modern day loading with a cat and grapple at Simpson Timber Company operations.*

*Aerial view of Humboldt Bay, looking north. The city of Eureka is on the right, with Gunther Island, once the site of the Excelsior Redwood Company mill, just beyond. On the left, the plumes of white smoke mark two paper mills, with the town of Samoa just beyond. Mad River Slough enters the bay beyond Samoa, while the city of Arcata marks the northern edge of the bay.*

# INDEX

(* Asterisk denotes illustration)

## A

Advance, 84*, 85*
Albion (town), 18, 34*, 35*, 59*, 67, 70, 105
Albion Lumber Co., 18, 58*, 66*, 72
Albion River, 16*, 70
Albion River RR., 58*, 66*, 70*, 72
Albion & Southeastern RR., 72, 115
Alco (American Locomotive Co.), 101*
Alta, 92*
Alton Junction, 113
Anaconda Copper Mining Co., 89*
Annie & Mary, 90, 99*
Antelope, 86*
Arcata, 83, 86, 90, 93, 94*, 95*, 99*, 105*, 141*
Arcata & Mad River RR., 90, 91*, 92*, 94*, 95*, 96*, 98*, 99*, 100*
Arcata Redwood Co., 128*
Arcata Transportation Co., 90
Astor Cut, 88*
Astor, John Jacob, 88*
Astoria, 92*
Astoria & Columbia River RR., 89*

## B

Baldwin (Locomotive Works), 56*, 57, 70, 97*, 100*, 107*, 109*, 110*, 114*
Bay Mill, 112*
Bayside, 106*, 107*
Bayside Lumber Co., 106*, 107*
Bayside Mill & Lumber Co., 107*
Big Blackfoot Milling Co., 89*
Big Bonanza Mill, 83*, 84*, 90
Big Lagoon, 130*, 131*, 132*, 133*, 137*, 139*
Big River, 13*, 14*, 16, 67*
Blackburn, Judge, 14
Black Diamond, 91*
Blue Lake, 94*, 95*, 96*, 97*
Bodega, 12, 16
Bodega Head, 33
Boise-Cascade Corp., 77, 79*, 80
Bonner, E. L., 89*
Bonner & Walsh, 89*
Booneville, 59*
Bourn's Landing, 36*
British Columbia, 42
Brookings, 123, 124, 125*, 126
Brookings Lumber & Box Co., 122
Brookings Timber & Lumber Co., 121*, 123
Brunswick, 37*

Bucksport, 107*
Bucksport & Elk River RR., 101*, 103, 105*, 107*

## C

California Iron Works, 50*
California & Northwestern RR., 113, 115
California & Oregon Lumber Co., 123, 124*, 126*
California Redwood Association, 128*
California Redwood Co., 103
California Western, 40, 76*, 77, 78*, 79*
California-Western Railroad & Navigation Co., 77
Carson, William, 103*, 112, 113*
Caspar, 37*, 38*, 56, 57
Caspar, 37*
Caspar Creek, 56, 68*
Caspar Creek RR., 56
Caspar & Hare Creek RR., 57
Caspar Lumber Co., 56, 68*, 69*, 71*, 73*
Caspar, South Fork & Eastern RR., 57, 72*, 73*, 74*, 75*
Chaffey, Bill, 20*
Chetco River, 121*, 123, 124
Civil War, 26
Cleone, 37*
Climax, 57, 73*
Columbia River, 39*, 51
Columbia River Packers Association, 89*
Comptche, 58*
Contra Costa County, 14
Cooper, James B. R., 11
Cooper's Molino Rancho, 6*
Crescent City, 120, 121*, 122, 128*
Crescent City & Smith River RR., 120, 122
Crescent City Wharf & Dock Co., 120
Cuffey's Cove, 67
Cullberg, Isaac, 20*

## D

Daisy, 71*
Dallas, Davidson & McPherson, 18
Delmar Landing, 53*
Del Norte Company, 123
Del Norte County, 94*, 120, 122, 124*
Del Norte Southern RR., 122
Dolbeer & Carson Lumber Co., 41, 99*, 103*, 105*, 110*, 112*, 114*

Dolbeer, John, 41, 45
Dollar, Captain Robert, 46, 47*
Dolly Varden Mill, 90, 94*, 105*

## E

East Bay, 12
Eddy & White, 14
Eel River, 112, 113, 114, 118*, 136*
Eel River & Eureka RR., 112, 114
Elk, 67
Elk Creek RR., 57*, 58*, 67
Elk River, 27*, 105*, 108*, 110*
Elk River Mill & Lumber Co., 109*
Elk River RR., 105*
Emily, 103
Essex, 90
Eureka, 10*, 11*, 16, 26*, 41, 83, 90, 92*, 93, 95*, 103*, 108*, 109*, 112*, 113*, 114*, 141*
Eureka Foundry Co., 106*
Eureka & Freshwater RR., 118*
Eureka Iron Works, 90
Eureka & Klamath River RR., 89*, 92*, 93, 113
Evans, David, 103*
Excelsior Redwood Co., 26*, 48*, 94, 103, 104*, 105*, 118*, 141*

## F

Falk, 105*, 108*, 109*
Falk, Noah H., 90, 94*, 105*
Father Crespi, 7
Finkbine-Guild Lumber Co., 67, 78*, 80*
Flanigan, Brosnan & Co., 106*, 107*
Forestville, 112
Fort Bragg, 15*, 17*, 37*, 40, 74*, 75*, 76*, 77, 79*
Fort Bragg RR., 71*, 74*
Fort Bragg Redwood Co., 77
Fort Bragg & Southeastern RR., 72, 115
Fort Humboldt, 10*, 109*
Fort Ross, 7
Freshwater, 26*, 104*, 105*, 119*
Freshwater Slough, 94

## G

Gannon Slough, 106*
General Sherman Tree, 3
Georgia-Pacific Corp., 89*, 130*, 131*, 136*
Glen Blair Redwood Co., 77*, 79*, 81*
Glendale, 95*, 96*
Globe, 108*, 110*, 122*

Glynn & Peterson Lumber Co., 53*
Goodyear Redwood Co., 67*, 68
Graham, Isaac, 11, 12
Greenwood, 59*
Greenwood Creek, 67
Greenwood Landing, 58*, 59*
Greenwood RR., 56*, 68
Gregg Expedition, 14
Gualala, 36*, 68
Gualala Mill Co., 54*, 55*, 56*, 59*
Gunther Island, 103, 104*, 141*
Gypsy, 85*, 87*, 92*, 97*

H

Hammond, A. B., 87*, 89*, 92*, 113, 114
Hammond Lumber Co., 89*, 128*,
   130*, 136*, 137*
Handly Chute, 34*
Happy Camp, 114*
Harbor, 123, 124
Hare Creek, 57
Harmon, S. H., 54*
Haywood, W. B., 54*
Healdsburg, 72
Heisler (Locomotive Works), 101*
Hercules, 73*
Hobbs, Wall & Co., 120, 121*, 122, 124,
   128
Holmes, 114*
Holmes-Eureka Lumber Co., 107*,
   138*
Hoopa, 97*
Hooper Brothers, 94, 100*, 103
Houda's Landing, 100*
Humboldt Bar, 33
Humboldt Bay, 14, 16, 30, 83, 94, 99*,
   100*, 103, 106*, 112*, 114, 115, 141*
Humboldt Bay & Eel River RR., 112
Humboldt Bay & Mad River RR., 93
Humboldt Bay & Trinidad Logging &
   Lumbering Co., 93
Humboldt County, 17*, 29*, 31*, 51*,
   83, 94, 105*, 108*, 109*, 112*, 136*
Humboldt Logging Ry., 94, 103, 104*
Humboldt Lumber Mills Co., 97*
Humboldt Northern RR., 110*
Humboldt State College, 94*
Humboldt Times, The, 30

I

Iran, S.S., 99*
Irmulco, 78*
Irvine & Muir, 78*

J

Jacoby Creek, 106*
Janes Creek Mill, 105*
Johnson, C. R., 15*, 79*
Johnson, Otis R., 79*
Jolly Giant Mill, 90, 91*, 105*
Jughandle Creek, 56, 57, 69*, 72*, 74*
Jumbo, 69*

K

Klamath River, 93
Korbel, 97*, 98*, 99*

L

Lake Earl, 120, 124
Lake States, 41
Lassen, Peter, 11

Lawson, Davenport, "Port", 67*
Lawson Skyline, 67*, 76*
Lewis & Clark Exposition, 110*
Libbey, Howard A., 128*
Lindsay Creek, 83*
Little River, 90*
Little River Redwood Co., 128*
Little Valley Lumber Co., 79*
Loggie, James, 111*
Los Angeles, 11

Mc

McCanse, Sam, 67*
McCanse Flyer, 67*
McCormick, Hugh, 20*
McKay, Allan, 111*
McKay & Co., 108*, 110*, 111*

M

Mad River, 24*, 42*, 44*, 83*, 85*, 90,
   92*, 94*, 95*
Mad River Slough, 86*, 90, 141*
Marin County, 7, 8
Mark West Creek, 6*, 11
Marsh Creek, 58*
Marshutz & Cantrell, 97*, 109*
Mary Ann, 93*
Mattole Landing, 37*
Mattole Lumber Co., 81
Mendocino, 9*, 18
Mendocino Coast, 16, 33, 36, 46, 67*,
   77, 79*, 81
Mendocino County, 56, 115
Mendocino Lumber Co., 13*, 14*, 18,
   58*, 59*, 67*, 69*, 70*, 79*
Mendocino RR., 67
Milford Mill & Lumber Co., 94, 103*
Mill Creek, 11, 15
Miller, Wayne, 130*
Miners Foundry & Machine Works, 53*
Minor and Falk, 90, 91*
Minor, Isaac, 90, 94*, 105*, 123
Minor Mill & Lumber Co., 95*, 96*, 98*
Mission Dolores, 7
Mission Gabriel, 11
Mission Santa Clara, 7
Mississippi, 78*
Missoula Mercantile Co., 89*
Montana, 87*
Monterey, 8
Mt. Hermon, 11
Muley Mill, 112*
Murphy, Albert Stanwood, 119*
Murphy, Simon Jones, 115*
Murphy, Stanwood A., 134*

N

Navarro Lumber Co., 59*
Newberg Mill Co., 29*
New England, 29, 41
Nip and Tuck Landing, 33*
Northern Redwood Lumber Co., 98*,
   99*
North Fork, 92*, 95*, 97*
Northern Pacific, 89*
Northwestern Pacific RR., 59*, 72, 115,
   122
Noyo, 17*, 39*
Noyo River, 40*, 57

O

Oakland, 12
Occidental Mill, 103*, 108*
Onward, 84*, 92*
Oregon & Eureka RR., 90*, 114
Oregon Pacific RR., 89*
Orr, Frank, 56*

P

Pacific Lumber Co., 102, 103, 104,
   106*, 107*, 108*, 109*
Palo Alto, 7
Park & Cliff House Ry., 53*
Peterson, A. F. "Bud", 136*
Pioneer Mill, 14
Point Arena, 33*, 34*, 58*, 67
Pope & Talbot, 89*
Porter, H. K., & Co., 89*, 111*
Portola Expedition, 7
Presidio of San Francisco, 7
Puget Sound, 51, 89*

R

Redwood Canyon, 12
Riverside, 102*
Riverside Lumber Co., 101*
Robertson Raft, 39*, 40*
Rockport, 39*, 57, 78*
Rogers (Locomotive Works), 117*
Rollerville Landing, 58*
Rowdy Creek, 124
Russian River, 6*, 11
Russians, 7, 11
Rutledge, Peter J., 114*
Ryan Creek, 108*
Ryan, Duff & Co., 10*, 14, 83*
Ryan Slough, 112*

S

Salmon Creek, 61, 94
Salmon Creek RR., 67
Salsig, 56*
Salsig, E. B., 58*
Samoa, 87*, 90*, 93, 141*
Samson, 74*
San Francisco, 7, 8, 12, 14, 53, 56, 59*,
   70, 97*, 105*, 108*, 110*
San Francisco Bay, 40*, 42, 113, 115
San Francisco & Northwestern RR.,
   114, 116
San Francisquito Creek, 7
San Lorenzo River, 12
Santa Clara, 14, 16, 83*
Santa Clara County, 8
Santa Cruz, 7, 8, 14
Santa Cruz County, 11
Santa Fe RR., 72, 113, 114, 115
Scotia, 112, 113, 115*, 116*
Sequoia, 65*, 100*
Shay, 81*, 120*, 126*
Simpson Timber Co., 90, 137*, 141*
South Bay RR. & Land Co., 93
Skunks, 40, 77, 79*, 81
Smilax, 72*
Smith River, 120, 124
Smith, Stephen, 12
Sonoma County, 11, 12, 53*
Sorenson, Henry, 81*
Southern Pacific RR., 72, 89*, 113, 115
Spanking Fury, 86
Stewart, Hunter & Johnson, 15*

Stitz Creek, 117*
Stout Tree, 2

T

Ten Mile River, 15*, 76*
Tiburon, 115
Trinidad (town), 15*, 94, 100*, 101
Trinidad Head, 14
Trinidad Mill Co., 94, 100*
Trinity County, 14
Trojan, 75*

U

Union, 86
Union Lumber Co., 15*, 75*, 76*, 77,
    79*, 81*

Union Plank Walk & RR. Co., 86, 91*
Union Wharf & Plank Walk Co., 86
Usal, 46

V

Vance, John, 83*, 90, 93, 112
Vance, John, Mill & Lumber Co., 21*,
    24*, 42*, 44*, 48*, 52*, 90
Vance's Mad River RR., 85*, 90
Van Duzen River, 113

W

Warren Creek, 90
Washington Iron Works, 134*
West Eureka, 87*
Westport Landing, 35*, 40*

Weston, F. E., 120
White, L. E., Lumber Co., 57*, 58*, 59*,
    67, 68
White, Martin, 112*
Whitmore, Jerry, 112*
Williams, Ted, 130*
Willits, 40, 56, 77, 78*, 113
Wonderly Mill, 46
Woodside, 7, 18

Y

Yosemite National Park, 94*

Z

Zaruba, V., 93*
Zayante Grant, 12

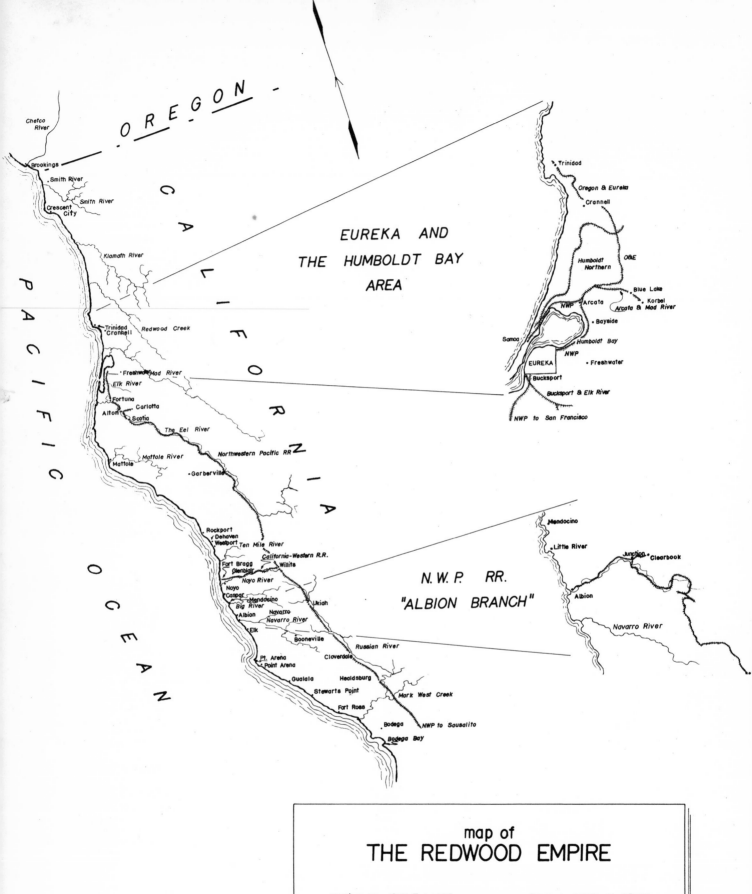

map of
# THE REDWOOD EMPIRE

compiled by   JOHN T. LABBE          Drawn by DAVID W. BRAUN

1974

GRAPHIC SCALE |‑‑‑‑‑‑‑| 20 miles